Date Due

JUL 3 1985		
JUN 1 9 1985 ℝ ℝ		
JUN 2 0 1985		

Demco 38-297

Also of Interest

The Biology of Social Insects: Proceedings of the Ninth Congress of the International Union for the Study of Social Insects, edited by Michael D. Breed, Charles D. Michener, and Howard E. Evans

Metabolic Aspects of Lipid Nutrition in Insects, edited by T. E. Mittler and R. H. Dadd

Orthopteran Mating Systems: Sexual Competition in a Diverse Group of Insects, edited by Darryl T. Gwynne and Glenn K. Morris

Pest Control: Cultural and Environmental Aspects, edited by David Pimentel and John H. Perkins

World Food, Pest Losses, and the Environment, edited by David Pimentel

Insect Behavior

About the Book and Editors

Insect Behavior: A Sourcebook
of Laboratory and Field Exercises
edited by Janice R. Matthews and Robert W. Matthews

Interest in insect behavior is growing rapidly, as reflected both in courses devoted fully to the topic and in its inclusion in general biology, ecology, invertebrate zoology, and animal behavior--as well as general entomology--curricula. Instructors and students find that insects are in many ways uniquely suitable animals for behavioral study: they are more approachable and predictable than vertebrates; one can experiment with them without public criticism and normally without special permits; they are available in large numbers; and they are inexpensive and relatively easy to maintain. However, few carefully planned exercises using insect examples have been published, and those available often are extremely difficult to use.

This sourcebook is a comprehensive, concept-oriented collection of 34 laboratory and field behavioral exercises using insects. The exercises, developed by 25 specialists, are original, creative, and illustrate fundamental principles of animal behavior. All have been classroom tested. Written in a modular format, they may be easily adapted to laboratory periods of various lengths. Most include suggestions for further study suitable for extra credit or individual research projects. All include selected references, summary questions, and data sheets where pertinent. Appendixes cover statistical procedures and offer "Aids to the Instructor," in which additional preparatory information is provided with hints, cautions, background, and where helpful, sample results. Descriptions of capture, culture, and handling methods are given, as well as suggestions for alternate experimental animals.

Janice R. Matthews, biomedical editor for the College of Veterinary Medicine at the University of Georgia, and Robert W. Matthews, professor in the University of Georgia Department of Entomology, have collaborated for many years in research on parasitic, solitary, and social wasps and have coauthored a widely used textbook, *Insect Behavior*.

Insect Behavior
A Sourcebook of Laboratory and Field Exercises

edited by Janice R. Matthews
and Robert W. Matthews

Westview Press / Boulder, Colorado

Copyright © 1982 by Westview Press, Inc.

Published in 1982 in the United States of America by
    Westview Press, Inc.
    5500 Central Avenue
    Boulder, Colorado 80301
    Frederick A. Praeger, President and Publisher

Library of Congress Catalog Card Number: 82-50918
ISBN: 0-86531-412-8

Composition for this book was provided by the editors
Printed and bound in the United States of America

Contents

x

Tables and Figures

Preface

Insects generally make far more suitable animals for behavioral study than do vertebrates. They are much more approachable. They are more predictable. One can experiment upon them without public criticism and normally without special permits. They are available in large numbers, making replication of observations much easier. They are less expensive and relatively easy to maintain.

The exercises included in this sourcebook represent the first offering designed to meet the need for laboratory and field experiences that deal comprehensively with insect behavior. They should fill a void that exists for college and university teachers and students, high school teachers and naturalists.

Our original request to the 25 contributors to this sourcebook was "Send us your 'winners' that work!" Illustration of important principles, exercise reliability, and student interest were prime criteria in selection of material. The contributors have developed and used their exercises in a variety of teaching situations. In many cases, we have also tested the activities ourselves.

In compiling this sourcebook, we have attempted to meet three needs. First is the need for laboratory and field activities to use in insect behavior courses. Exercises have been solicited to cover most major areas included in a well-balanced laboratory course in insect behavior. Both traditional and non-traditional exercises have been included. The species selected have been those from which our contributors have obtained satisfactory results under normal teaching conditions. A wide variety of activities have been sought, with a suitable diversification between laboratory and field work, and with a seasonal and geographic balance within the latter. Most of the species used are readily available, but the exercises can also serve as models for similar work with other organisms.

Second is the need for exercises which, while using insects as experimental subjects, impart fundamental ideas of contemporary behavioral study. The activities presented here may be easily incorporated into other biology courses for enrichment or as substitutes for activities using more expensive and/or less available animals. We have attempted to focus on concepts and deemphasize terminology. An entomological education is not a prerequisite for understanding these exercises. Quantitative procedures are included where appropriate, but only a knowledge of basic algebra is assumed.

Third, we recognize that many behavior courses include independent work. Thus, most exercises include Suggestions for Further Study to propose activities and ideas which might serve as the basis for individual research projects. The sourcebook begins with pointers for individual research projects and reports, and concludes with appendices of use to independent researchers.

After an introduction which briefly describes the background for the problem, each exercise is presented in a standard format under the following headings:

Subject -- a brief description of the preferred insect species, often with suggestions for alternative organisms.

Materials Used -- a shopping list of all equipment and supplies required, nearly always of a simple kind.

Procedure -- a detailed description of how the exercise is to be conducted.

Each exercise concludes with summary questions and a pertinent bibliography. A separate section, Aids to the Instructor, provides extensive pedagogic assistance for each exercise, including detailed prelaboratory preparation information, descriptions of culture and/or handling methods, hints, cautions, additional background, and, where helpful, sample results. The first two appendices provide an introduction to commonly used nonparametric statistical tests, detailed examples of their application to problems similar to those in this sourcebook, and relevant tables. A third appendix points out sources of supplemental materials, including information about where to obtain cultures of most species used in these exercises.

The preparation of this manual has been greatly facilitated by the willing and timely cooperation of the contributors, all of whom are outstanding researchers and teachers. We are grateful to them all. This book has been prepared for teachers and students; comments by users will be welcomed and should be directed to the authors of particular exercises, who alone are responsible for the content and development of their materials.

We would like to express special gratitude to the Division of Neurobiology and Behavior, Cornell University, for providing a stimulating and pleasant environment in which to spend the sabbatical year during which this project was conceived. The University of Georgia provided support, both secretarial and financial, without which this project could not have been executed. Students in the insect behavior course at the University of Georgia have been keen "guinea pigs"; in innumerable ways over the years they have helped develop and improve several of the exercises. Many people have provided help and encouragement along the way; in particular, we would like to thank Christine Boake, W. L. Brown, Jr., George C. Eickwort, Preston Hunter, and Maurice Tauber. Very competent secretarial assistance has been provided by Deborah Anderson, Bertha Blaker, and Terri Natoli.

Athens, Georgia

Janice R. Matthews
Robert W. Matthews

Contributors

Roger D. Akre, Department of Entomology, Washington State University, Pullman, WA 99164

John Alcock, Department of Zoology, Arizona State University, Tempe, AZ 85281

Edward M. Barrows, Department of Biology, Georgetown University, Washington, DC 20057

Paul D. Bell, Department of Zoology, Erindale College, University of Toronto, Mississauga, Ontario L5L IC6, CANADA

William H. Cade, Department of Biological Sciences, Brock University, St. Catharines, Ontario L2S 3A1,, CANADA

Philip S. Callahan, Insect Attractants, Behavior and Basic Biology Research Laboratory, USDA-SEA, P.O. Box 14565, Gainesville, FL 32604

William G. Eberhard, Escuela de Biología, Universidad de Costa Rica, Ciudad Universitaria, San José, COSTA RICA

Lee Ehrman, Department of Biology, State University of New York, Purchase, NY 10577

George C. Eickwort, Department of Entomology, Cornell University, Ithaca, NY 14850

T. D. Fitzgerald, Department of Biological Sciences, State University of New York, Cortland, NY 13045

Robert Franklin, Department of Biology, University of Oregon, Eugene, OR 97403

William H. Gotwald, Jr., Department of Biology, Utica College of Syracuse University, Utica, NY 13502

Bernd Heinrich, Department of Zoology, University of Vermont, Burlington, VT 05405

Rudolf Jander, Department of Entomology, University of Kansas, Lawrence, KA 66044

Janice R. Matthews, Department of Entomology, University of Georgia, Athens, GA 30602

Robert W. Matthews, Department of Entomology, University of Georgia, Athens, GA 30602

Glenn K. Morris, Department of Zoology, Erindale College, University of Toronto, Mississauga, Ontario L5L IC6, CANADA

Lowell R. Nault, Department of Entomology, Ohio State University, Wooster, OH 44691

William L. Nutting, Department of Entomology, University of Arizona, Tucson, AZ 85721

Ira B. Perelle, Biology Department, Mercy College, Dobbs Ferry, NY 10522

Ronald L. Rutowski, Department of Zoology, Arizona State University, Tempe, AZ 85281

Robert L. Smith, Department of Entomology, University of Arizona, Tucson, AZ 85721

John G. Stoffolano, Jr., Department of Entomology, University of Massachusetts, Amherst, MA 01003

R. Stimson Wilcox, Department of Biology, State University of New York, Binghamton, NY 13902

Chih-Ming Yin, Department of Entomology, University of Massachusetts, Amherst, MA 01003

OBSERVATION, DESCRIPTION, AND ANALYSIS
OF BEHAVIOR

1. Conducting a Personal Project on Insect Behavior in the Field

John Alcock
Arizona State University

SELECTING AND CARRYING OUT A PROJECT

Choosing a subject to observe is a crucial task. It is easy to feel a certain amount of uncertainty about picking something to study. Will you find anything at all to observe? Will it be worth studying if you find it? There are a number of simple rules that may help to answer these questions.

1. Get out and look around.

Insects are everywhere, even inside your homes and classrooms, but you will probably have the most success looking for potential research topics in vacant lots, city parks, and the countryside (if you are fortunate enough to have access to rural habitats). I began my career as an insect ethologist on a camping trip that took me to a heavily grazed scrub pasture which was populated by cows and large numbers of a big black, green, and orange grasshopper. The grasshoppers proved interesting and easy to study. Upon returning to my home in Seattle, I examined a nearby vacant lot overrun with blackberry bushes which I discovered were inhabited by an immense population of stinkbugs. My study of these animals was just as rewarding to me as my work with the grasshoppers. And the more time I spent in the vacant lot, the more intriguing insects I found. In 2 summers I was able to complete only a fraction of the possible research projects available at this suburban site. There were larvae of a species of ground beetle collecting grass seeds and carrying them into an underground burrow, 11 species of digger wasps that nested along a little path through the lot, leafcutter bees that carried the leaf-linings for their brood cells into underground chambers, damselflies that mated in a swampy portion of the lot, and long-horned grasshopper males that could be watched singing their songs in the fall and attracting females to them, to name just a few of the thoroughly urban insects available for watching. I am confident that there are few places in the world, even in large cities, where observable insects cannot be found at some time of the year. If additional inspiration is needed, let me recommend Howard Evans' Wasp Farm and Life on a Little-Known Planet.

1

2. <u>Work with a common insect that is reliably present in substantial numbers in one or more locations</u>.

Only if you are able to find your animal repeatedly over a considerable period of time will you succeed in acquiring sufficient data to prepare an interesting and complete report.

3. <u>Develop a set of questions that you can answer in the time available</u>.

If what you see does not stimulate your curiosity then do not study this insect species. Remember that the point of the exercise is to permit you to study something because you wish to satisfy your curiosity about an insect's behavior. If this element of motivation is absent, you will probably be bored by your research and will do a mediocre job.

4. <u>Establish a set of priorities</u>.

If you have developed a list of 6 answerable questions, you must decide which of these is the most important, which ranks second, and so on. <u>Answer one question at a time</u>, collecting <u>all</u> the information and quantitative data that are required to solve this problem before moving on to the next question.
It may be possible at times to gather data on 2 or even 3 questions more or less simultaneously. Do not be tempted to do so. I cannot stress too strongly the need to be disciplined and to restrict the scope of your observations. It is all too easy to dissipate one's time and energy in a futile effort to collect material on many topics at once. The attempt to do too much at the same time is a recipe for disaster. It will almost invariably lead to a shallow, inconclusive report in which a little bit is said about a wide variety of issues. It is better to answer 1 or 2 questions thoroughly than to pose 6 or 7 questions and answer none of them to your satisfaction.

5. <u>Collect numerical data</u>.

Whenever possible, pose your questions in such a way that they can be answered quantitatively. Numbers have the great advantages of concreteness and precision. They enable another person to assess the reliability of the observer's conclusions. If it is appropriate they can be subjected to statistical analysis, provided the data were properly gathered. It is far more satisfactory to be able to present a graph based on actual counts of mating pairs at regular intervals through the day that shows a sharp peak between 3 - 6 p.m. than to assert, without supporting evidence, that your insects tended to mate in the late afternoon. Lehner (1979) provides a thorough discussion of such things as methods of data collection, statistical analysis, and graphical presentation. Colgan (1978) discusses quantitative techniques for the advanced student.

6. Don't jump to conclusions.

Haste in interpreting the function of what you see your animals doing can bias what you do. For example, quickly labeling an activity an "attack response" might prevent you from seeing that it was really male courtship activity directed to a female. Try to give the behavioral components of your insect neutral labels, and keep an open mind about their function until you have collected enough information to be confident of your conclusions.

7. Keep things simple.

You'd be amazed at how much you can do with a pencil, a notebook or data sheet, a stopwatch or wristwatch, and a supply of patience. There are many more sophisticated tools of field research (see Lehner, 1979) but you are unlikely to need them to carry out a project successfully.

If you have a single lens reflex camera and are already an insect photographer, you may wish to devote some time specifically to photographing your subject's behavior. Don't plan to combine data collection and still photography. The 2 activities require attention to different things.

There are many super-8 movie cameras that require no experience to operate successfully. Films can be useful if you wish to describe or analyze a complex or rapidly performed activity, or if you wish to make a film to illustrate your oral presentation (see Dewsbury, 1975).

GIVING AN ORAL REPORT

Some elementary advice on how to give an oral presentation may be useful. I shall assume that you will have only 5 - 15 minutes in which to present your findings, a short time in which to discuss the results of several weeks of research.

1. Don't rush.

In any talk, especially a short one, you will be able to convey only a few major points. Accept this restriction gratefully, and do not try to hurry through a large number of topics, major and minor. A self-imposed limitation on the number of issues to discuss benefits both the speaker and listeners. The members of any audience have the capacity to absorb only so much information. Do not overload their circuits or they will tune you out.

2. Organize your talk carefully.

Make it revolve around the 2 or 3 points you wish to make. Do not leap back and forth, touching first on 1 issue, then on a second, then back to the first again. At the very outset of your presentation, let your audience know the structure of your talk, and the issues that you are going to be discussing. (An old recipe for a successful talk is: "Tell 'em what you're gonna tell 'em; then tell 'em; then tell 'em what you've just told 'em.")

3. <u>Remember your audience</u>.

Do not assume that your audience possesses special background information on your topic. It is difficult for a person who has lived with a research problem for some time to put himself in the position of a listener totally unfamiliar with this subject. Try to remember what you knew at the start of your project and adjust your talk accordingly.

4. <u>Present your results in simple, direct ways</u>.

Use slides, charts, or tables. Visual aids should be visible from the back of the room. Each aid should make one point only. Do not prepare hypercomplicated illustrations that summarize all your research in a mass of data (Fig. 1-1). Your audience will not make the effort to decipher your results, and you will lose them.

5. <u>Convey honest enthusiasm.</u>

It is hard for listeners to become excited about a topic if the speaker himself appears uninterested. Most of you will be moderately nervous before giving your talk. Perhaps it will help to know that your feelings are nearly universal, and that they tend to disappear once you actually begin to speak.

WRITING A FINAL REPORT

Writing a paper on your independent project does not differ substantially from preparing any scientific report. Much more emphasis, however, will need to be placed on the "introduction" and "materials and methods" than in laboratory reports on class work where these matters are largely described in the manual itself. Carry out a thorough search of the literature using the <u>Zoological Record</u>, <u>Biological Abstracts</u>, and the <u>Source</u>, <u>Citation</u>, and <u>Subject Indexes</u>, among other references, to establish what is already known about your subject. When you find a useful reference, you may be able to use it to initiate the "chain reaction method" of searching the literature by looking up papers cited in this article to see what citations they, in turn, may yield.

There is no 1 way to integrate your observations in your report. Read a variety of original papers on topics related to your research, and study the most clear and useful ones for ideas about how to organize your paper. In your written report, as in your oral report, a key to success will be your ability to focus attention on a limited number of major messages that you wish to convey. This requires careful thought about clarity and conciseness in writing style and inventiveness in the presentation of your results and conclusions. Proper grammar and correct spelling are <u>not</u> trivial matters. Errors in spelling and grammar distract and confuse a reader, diluting the impact of your report. I encourage you to read a little book, <u>The Elements of Style</u> by William Strunk and E. B. White; its advice

Fig. 1-1. A comparison of 2 ways of presenting the same information. The tabular presentation attempts to summarize too much data. The lower figure is much easier for the audience to grasp and presents sufficient information in a pleasing and uncomplicated manner.

A COMPLETE CATALOG OF THE SEQUENCE OF COURTSHIP ACTIVITIES OF MALE AND FEMALE LARGOBUGGA BIFURCATA OBSERVED IN BELLEVILLE, AND SURROUNDING AREAS DURING MAY 1980 and JUNE 1981.

1.	Male approach female	N = 47	5a.	Female departs	N = 9
2.	Male taps female	N = 41	5b.	Female turns	N = 19
3a.	Female retreats	N = 13	6a.	Male buzzes wings	N = 13
3b.	Female waves legs	N = 28	6b.	Male taps female	N = 2
4a.	Male taps female	N = 2	6c.	No response by male	N = 4
4b.	Male bumps female	N = 20	7.	Female waves legs	N = 5
4c.	Male departs	N = 3	8.	Male mounts female	N = 15
4d.	Male buzzes wings	N = 3	9.	Copulation	N = 12

Typical Successful Courtship of the Large Bug

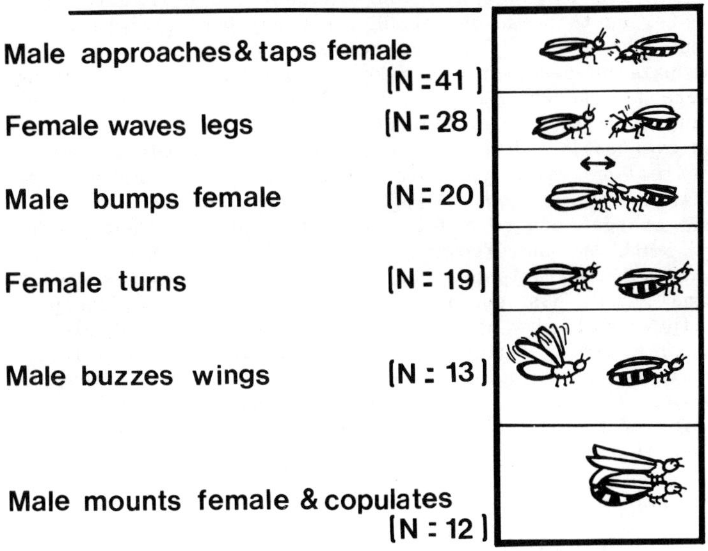

Male approaches & taps female (N = 41)

Female waves legs (N = 28)

Male bumps female (N = 20)

Female turns (N = 19)

Male buzzes wings (N = 13)

Male mounts female & copulates (N = 12)

is particularly relevant to writers of scientific reports, who sometimes forget that the object of the exercise is communication. One tactic that sometimes helps produce a more readable paper is to read aloud your first draft. If your writing sounds awkward, <u>revise</u>.

PREPARING A PAPER FOR PUBLICATION

There is no reason why an ambitious and intelligent student cannot do a special project of sufficient excellence to merit publication in a scientific journal. Many students have done work of publishable quality in animal behavior courses, and some have succeeded in producing a short paper based on their class project. Students often feel that anything they could do would surely have already been studied and would already appear in the literature. My own experience is relevant to this concern. After I completed my field work on the grasshopper mentioned earlier, I began a search to discover what had been published previously on this animal. I assumed that such an interesting, exceptionally conspicuous, and abundant creature would have attracted considerable attention. It seemed likely that my findings had already been published by an earlier observer. To my great surprise, I learned that practically nothing had ever been written about this grasshopper. Subsequently I have come to be surprised if I <u>do</u> find something in the literature on the behavior of an insect species that I have studied. The natural history of only a tiny fraction of the hundreds of thousands of insect species is known. Even the behavior of the few heavily studied species is still imperfectly understood. For example, the honey bee is without doubt the most intensively studied insect. Yet in recent years literally hundreds of papers have appeared on various aspects of its biology, and you can be certain that there are many more mysteries to be solved about the honey bee's behavior.

Thus the field of insect behavior is an open one in which a careful patient observer can make a contribution, discovering things that were not known before. I would like to encourage you to aim high, and at least to consider the possibility of doing research that you can eventually publish. Consult your instructor on the selection of a journal to which the work can be submitted, on the form of the manuscript, and on how to cope with page charges, if any. (Many journals charge authors a fee for each page published, although for student authors the fee is often waived if the student lacks institutional or grant support.) Generally one sends a manuscript to a journal that has published similar studies in the past. The manuscript should be as concise as possible and should precisely follow the format employed by the journal. In addition, initial drafts of the paper should be reviewed not only by the instructor but by several other qualified people, and revised as necessary prior to submitting a final draft to a journal. Publishing a paper is hard work, but it is also a great thrill and I recommend it highly.

SELECTED REFERENCES

Colgan, P. 1978. <u>Quantitative Ethology</u>. John Wiley and Sons, New York.

Dewsbury, G. D. 1975. Filming animal behavior. In E. O. Price and
    A. W. Stokes (eds.). <u>Animal Behavior in Laboratory and Field</u>,
    2nd ed. W. H. Freeman, San Francisco, pp. 13-15.
Evans, H. E. 1963. <u>Wasp Farm</u>. Natural History Press, New York.
Evans, H. E. 1966. <u>Life on a Little-Known Planet</u>. Dell, New York.
Lehner, P. N. 1979. <u>Handbook of Ethological Methods</u>. Garland STPM,
    New York.
Strunk, W. Jr. and E. B. White. 1972. <u>The Elements of Style</u>. 2nd.
    ed. MacMillan, New York.

2. Observation, Description, and Quantification of Behavior: A Study of Praying Mantids

Edward M. Barrows
Georgetown University

Detailed studies of animal behavior usually proceed through 3 phases: (1) observing and describing behavior, (2) quantifying behavior, and (3) relating one's findings to already existing principles to substantiate or modify these principles or to discover new principles.

"Just watching" comes first -- observing, and preparing a list of specific behaviors to define an animal's behavioral repertoire. Many descriptive studies are concerned only with compiling such a behavioral catalog (ethogram), and appear to make only minimal use of explicitly quantitative data. (However, description is itself a quantitative process, though the quantification may be done at an intuitive and perhaps partly unconscious level.)

As one attempts to discover possible correlations that are more elaborate and/or remote from the original observations, quantitative methods play an increasingly larger role. Experimental work requires particularly accurate observation, description, and quantification. To insure that data are collected in the ways best suited for answering research questions, a well-designed experimental protocol with quantitative analysis should be planned before the relevant data are collected. (However, you should feel free to modify the protocol as you learn more about the animals' behavior, and as your question becomes more well defined.) Furthermore, it is important that the exact sampling procedure used in a study be clearly stated.

In this exercise, we shall study praying mantids as an introduction to some aspects of the observation, description, and quantification of insect behavior. Specifically, we shall examine feeding, grooming, defensive behavior, mating, and cannibalism. (Suggestions for Further Study include egg-laying and hatching.) In the process, you will have the opportunity to gain experience in determination of behavioral units, experimental manipulation, analysis of behavioral patterns, ethogram construction, and the calculation and use of certain descriptive statistics and 2 commonly used inferential statistics, the Mann-Whitney U Test and the Fisher Exact Probability Test.

METHODS

Subject

There are about 2,000 species of Mantidae worldwide. Common species in the continental USA include Mantis religiosa L. (from Europe), Stagmomantis carolina (L.) (native), Tenodera ardifolia sinensis Saussure (from China), and Brunneria borealis Scudder (native). Evidentally, Brunneria borealis has only parthenogenetic females (Borror et al., 1981), making it unsuitable for observing mating. Otherwise, any common species may be used in this exercise.

Praying mantids are diurnal insects that tend to be quite active under classroom conditions; a warm (about 26°C), bright classroom encourages their activity. Adult male mantids tend to have larger (and possibly more sensitive) compound eyes than females; adult females are usually larger and physically much stronger. Adults of both sexes will mate 10 days or longer after their final molts; females that are ready to oviposit are unlikely to mate. Various aspects of the mantid life cycle are shown in Fig. 2-1.

Materials Needed

Minimum of 5 male and 5 female adult mantids of the same species; 3 crickets for every mantid used in feeding demonstration; terraria (5 or 10 gallon) with clear glass sides and glass or plexiglas tops; 2 spray bottles that produce a fine mist; 2% acetic acid solution; leafy branches that fit into the terraria (privet, Ligustrum, works well); white poster board "arena", 40 cm high and large enough to encircle a terrarium; small scissors; graph paper; calculator with a standard deviation program, or statistics text that shows how to compute standard deviations; stopwatches. Desirable: standard biostatistics textbook such as Zar (1974) or Sokal and Rohlf (1969). Optional, for Further Study Suggestions: movie equipment with a closeup lens, or videotaping equipment with a closeup lens. (If immature mantids are used, they should be fed appropriately-sized prey such as Drosophila or immature crickets.)

Procedure

A. Defense: Defining the Units of Behavior

One of the key procedures in ethological studies is compilation of an ethogram, or behavioral catalog. It is based upon a determination of individual behavioral units, and may vary according to the way in which these are defined. What will be the basic behavioral units for a defense catalog for mantids? How will they be defined? (We cannot measure what we cannot define.)

In general terms, anti-predator or defensive behavior in mantids has been said to be comprised of primary and secondary patterns (Robinson, 1969a,b). Primary defensive adaptations reduce the probability of a predator's initiating an attack. They include

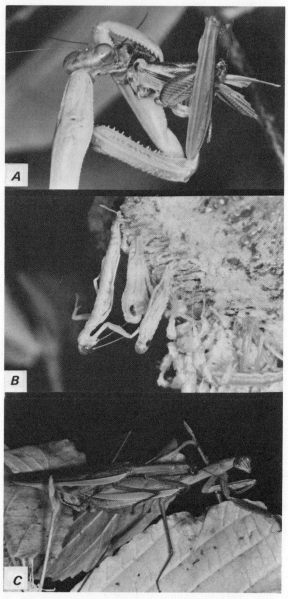

Fig. 2-1. _Tenodera_ _ardifolia_ _sinensis_: (A) feeding on a house cricket, _Acheta_ _domesticus_; (B) first instars emerging from their ootheca; (C) copulating.

remaining motionless, selecting perches where mantids are camouflaged, and changing color to match backgrounds. Secondary defensive adaptations are shown after a predator attacks, and act to reduce the probability of a successful attack. They include feigning death (thanatosis), active escape, startle (deimatic) display, stridulation that may resemble a hissing snake, and aggressive "retaliation". Within these broad groupings, of course, more finely drawn categories are nested.

Working in pairs or teams, obtain a mantid in a terrarium. As 1 person provides the stimulatory input, the other should describe the behavior in plain, objective language in chronological order in a notebook. Use Fig. 2-2 as an aid to accurate description of whatever body parts are involved.

Move your hand close to the mantid in the terrarium. How does the mantid respond to this movement? Describe as specifically as possible. Pick up the mantid by its thorax. How does it respond now? After you release it? Place the mantid in an empty arena. Repeatedly touch the mantid's side with your hand. How does it respond to this?

Consider what you have written. Does the stream of behavior divide naturally and easily into separate, distinct units of behavior? As a group, attempt to define and describe behavior units for further study of this aspect of defense. In order to elicit defensive behavior, you had to provide a "threatening" stimulus. Discuss what you will use as a standard. (How important does standardization of the stimulus appear to be?)

Select a single defense behavior that occurs frequently but is too fast to measure accurately with a manually-operated stopwatch, such as a strike. While presenting your "threat" at predetermined intervals, observe a single individual for 10 consecutive min. For each time interval (elapsed) record the number of times that the selected behavior is performed. You are measuring the rate of occurrence of an event.

Select another behavior that is relatively long-lasting, one that can be measured or timed, such as remaining motionless. Over a 10-min. observation period, record the total amount of time that your subject spends engaged in the behavior selected. Now you are measuring duration of a state.

Compare and contrast the results of the 2 approaches to obtaining quantitative information about behavior. Suggest other methods by which quantitative data regarding the mantid's defensive behavior could be obtained.

B. Grooming: Looking at Behavioral Patterns

In cataloging an animal's behavior, you may find that certain behaviors appear to occur in some sort of order. Such sequences may be described by simply listing the acts in succession, or you may tabulate the starting times of each act along with the name of the act itself. There are methods of statistically analyzing either case.

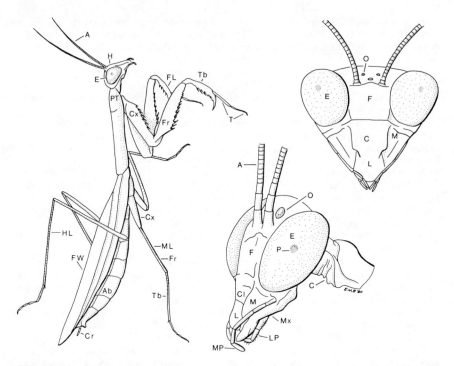

Fig. 2-2. External anatomy of a mantid, <u>Tenodera</u> <u>ardifolia</u> <u>sinensis</u>.
A, antenna; Ab, abdomen; C, cervix (neck); Cl, clypeus; Cr, cercus;
Cx, coxa; E, compound eye; F, frons; FL foreleg; Fr, femur; FW,
forewing (tegmen); H, head; HL, hindleg; L, labrum; LP, labial palp;
M, mandible; MP, maxillary palp; O, ocellus; P, "pupil"; Pt,
prothorax; T, tarsus; Tb, tibia.

    Mantids, like many other arthropods, instinctively clean each
part of their bodies, using both their legs and mouthparts to groom
themselves. Does mantid grooming show a particular sequence? How
variable is it?

    As before, the first step is definition of the units of
behavior, and this may be done at several levels. For example, in
broad terms mantid grooming has been shown to include:

    <u>Foreleg grooming</u>: bringing the foreleg and mouthparts
together and mouthing (palpating) the foreleg. Mantids
often show this behavior when pieces of prey are stuck
between foreleg spines.

Antenna grooming: grasping the antenna with the
ipsilateral foreleg (the one on the same side of the body
as the antenna) and bringing it to the mouthparts, which
then grasp the antenna. The foreleg releases the antenna,
which is then mouthed. Relatives of mantids, including
cockroaches and crickets, groom similarly.
Head grooming: repeatedly moving the foreleg across the
head, sweeping successive areas of the head.
Coxal base grooming: abducting the forelegs and bending
the head ventrally towards foreleg coxal bases, which are
mouthed.
Walking leg grooming: grasping a middle or hind walking
leg by the ipsilateral foreleg, and bringing it to the
mouthparts for mouthing. Usually only the tarsus of the
walking leg is cleaned.
Body grooming: scratching with a walking leg towards the
posterior of the body.

Each of these units can be further subdivided. For example,
Zack (1978a,b) did a detailed quantitative study of S. lineola head
grooming. Usually it consists of from 4 to 6 cycles of movements
made on the same side of the head. Each cycle involves:

(1) foreleg protraction and head rotation that brings the
femur brush (Fig. 2-3A) in contact with the head;
(2) foreleg retraction and head rotation that moves the
femur brush along the surface of the head; and
(3) femur brush cleaning with the mantid's mouthparts.

Each successive cycle cleans a more anterior and medial portion of
the head (Fig. 2-3B,C).
To determine whether your species of mantid shows a predictable
sequence in grooming its body parts, each team of students should
obtain a mantid. With a spray bottle that makes a fine mist, 1
person should spray the mantid with lukewarm water. Attempt to cover
all of its body. The other person should begin timing immediately,
and continue to do so for 15 min.
Time how long it takes the mantid to start grooming, and how
long it continues to groom after the spraying. Record how many
times, and in what order, a mantid grooms each part of its body. To
begin with, keep your behavioral units broad; if time permits, you
may wish later to study grooming of a particular body part in more
detail.
What is the total grooming frequency? What is the frequency of
grooming of each part? Does a mantid groom certain parts more than
others? What proportion of total grooming time is spent on each type
of grooming? Does there appear to be any order or sequence to the
grooming actions?
Repeat with the same mantid, using a lukewarm 2% acetic acid
solution to release grooming. As before, observe and time mantid

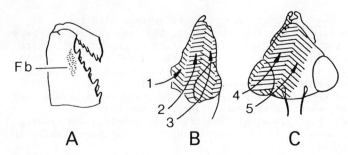

Fb

1
2
3
4
5

A   B   C

Fig. 2-3. Mantid grooming, showing femur brush (Fb) on foreleg (A), and lateral (B) and frontal (C) views of mantid head with five different areas that the foreleg cleans. The arrows show the approximate course of the femur brush as it cleans the head. (Redrawn from Zack, 1978a.)

behavior for 15 min total. Do your results appear to vary from those using lukewarm water? Pool class data and discuss your results.

C. Feeding

1) The Use of Descriptive Statistics to Organize and Summarize Data

About 1 hour before class, your instructor placed a mantid that had not been fed for 24 hr into a terrarium with some leafy branches. The well illuminated terrarium was encircled with an "arena" to eliminate unnecessary visual stimuli.

Place 3 crickets in the terrarium for large mantids, or about 10 Drosophila adults for small mantids. With a stopwatch immediately begin timing: (1) how long it takes until the mantid begins to orient visually toward the moving prey, and (2) how long the mantid feeds on a prey item after capturing it. Record what part of the prey is eaten first, second, and so on. Fig. 2-1A shows a mantid eating a cricket. Males often do not eat entire crickets.

Do individual mantids show differences in the sequences of prey parts that they eat? Does a mantid drop edible prey parts? If so, what are they? How does a mantid stop prey from struggling? Does a mantid kill its prey before eating it? How efficient is a mantid in prey capture? Does a mantid ever miss a prey during a strike of its forelegs? Do mantids catch and hold more than 1 prey at the same time? If a mantid grooms while eating or after eating, what parts of its body does it clean?

After collecting data, a behaviorist generally looks at the information in 1 of 2 ways. The first is through descriptive statistics, which help organize and summarize data. The second is through inferential statistics, which help test hypotheses and generalize from the data. Common descriptive sample statistics include these measures, calculated as follows:

<u>sample mean</u> ($\bar{x}$): sum the sample data (x_i, each score, measurement, or observation), and divide by the sample size (N, the total number of observations obtained).

$$\bar{x} = \sum_{i=1}^{n} x_i N$$

<u>sample median</u> (M): arrange the data in increasing order of magnitude, and find the value with an equal number of scores on both sides of it. If there is an even number of data, there will be 2 middle measurements, and the median is the mean of these 2. A difference between median and sample mean demonstrates that the sample data are not normally distributed.

<u>sample range</u>: the difference between the highest and lowest measurements in a group of data. The sample range serves as an estimate of the dispersion of the values in the population, but is a relatively crude and poor one, since a given sample is unlikely to really contain both the highest and lowest population values.

<u>standard deviation</u> (SD): when means are compared for data samples from 2 populations, it is also important to know how much variability there is in the original measurements from which those means were derived. The standard deviation is a measure of that variability about the mean. Standard deviations are calculated as indicated in Sokal and Rohlf (1969) or other standard statistics texts, or with a calculator with a built-in SD program.

$$SD = \left[\sum_{i=1}^{n} (x_i - \bar{x})^2 / (N - 1) \right]^{\frac{1}{2}}$$

<u>standard error of the mean</u> (SE): the standard deviation of the entire sample is divided by the square root of the sample size. This is a valuable measure because with the appropriate tables it can be used to calculate a range around the sample mean in which one feels confident the population mean lies.

$$SE = SD/\sqrt{N}$$

Use these descriptive measures to characterize your pooled class data on mantid feeding. For times recorded for each particular activity, ascertain the mean, standard error of the mean, and range. Do your sample data appear to be normally distributed?

If time permits, you may also wish to prepare a histogram to give a visual image of skewness and location for the data. Choose a standard interval length for the horizontal axis. Record the frequency (number of occurrences) of your measurements for each interval on the vertical axis. Plot these on graph paper, indicating the median and mean.

2) The Use of Inferential Statistics: The Mann-Whitney U Test

Very commonly one wishes to compare 2 samples to infer whether the populations they represent are statistically different. For a valid answer, some statistical methods (called "parametric") require that certain assumptions be met -- a normally distributed population, homogeneity of variance, and use of an interval or ratio measurement scale. Others, called nonparametric methods, do not make these demands. Their name comes from the fact that they draw inferences about populations without making certain assumptions about statistical population variables called parameters, e.g., the mean and SD.

Nonparametric tests are in general less "powerful" than parametric ones. That is, they have a lower probability that you will make a correct decision in your favor in interpreting the data. They are, however, relatively quick and easy to perform, and can be used when you are not sure whether your data meet the parametric assumptions (or are sure they do not).

One of the most commonly employed and most powerful nonparametric methods is the Mann-Whitney U Test. This test uses the ranks of the measurements instead of the actual measurements of variability of the 2 populations you are comparing.

The Mann-Whitney U Test can be used to compare feeding rates between male and female mantids, using your pooled class data. Follow the stepwise procedure given in Appendix 1.4. If you need further help, consult the hypothetical example given in Table 2-1.

Note: This and most statistical tests require independent and random data points to yield mathematically valid results.

D. Mating Behavior and Cannibalism: (The Fisher Exact Probability Test)

Introduce a male mantid into each terrarium containing a well-fed female. Remain motionless and observe. With a stopwatch, time how long it takes a female to track a male visually, and how long it takes a male to track a female. How long does it take a male to approach, mount, and start copulating with a female? Does the male show any courtship behavior (= behavior that may lead to copulation) with his antennae or other body parts? How might you design experiments that test whether or not males use pheromones, movement, size, shape, or color of females in locating them?

A male mounts and grasps a female with his foretibiae fitting into small grooves on the anterior-ventral part of her mesothorax just in front of her wing bases. His tarsi on his middle and hind legs are on the edges of the female's wings. How does his mounting affect female behavior?

Did your female attempt to eat the male during mating? Male cannibalism by females is not the usual behavior in mantids, despite what some textbooks might lead us to believe. Its extent appears to vary with the species, degree of disruption (artificial conditions,

TABLE 2-1. Hypothetical example of Mann-Whitney U Test used to compare times that male and female mantids take to feed on adult crickets. Mantids were timed from initiation of feeding until they either
completely ate a cricket or dropped all remaining parts of it.

The experimental hypothesis was that males feed more slowly than
do females. If males (in the entire population of mantids) do not
really take more time to feed than do females, the probability that
these data yield a U of 99 by chance alone is less than 1 out of 1000
(p < 0.001, 1-tailed test). Therefore, males may be said to feed
statistically slower than females in this hypothetical experiment.
Males also show more variability in their feeding times (SE = 38.58)
than do females (SE = 19.41).

| Individual | Feeding Time in Seconds | | | |
Number	Males	(Rank)	Females	(Rank)
1	1100	14	900	7
2	1350	20	941	9
3	955	10	859	4
4	1214	18	785	1
5	1193	17	913	8
6	1096	13	798	2
7	1293	19	887	5
8	1154	15	818	3
9	1188	16	972	11
10	999	12	896	6
Mean	1154.2		876.9	
Median	1171		891.5	
SD	121.98		61.39	
SE	38.58		19.41	
Range	955 - 1350		785 - 972	

disturbances, or irregularity in normal mating sequence, even the
mere presence of an observer), and length of time since the female
last ate, mated, or both.

Assuming that we could keep the other factors constant, let us
consider possible species differences in degree of cannibalism. In
biostatistical language, we are asking: is the possible difference
observed between species likely to occur infrequently (e.g., 5 times
out of 100, p = 0.05) by chance alone? Note: We are comparing only
cannibalism or no cannibalism here, not more than 2 classes of information. We assume random and independent sampling.

You may have encountered the chi-square statistic (Appendix 1.1,
1.2): it is the most common means of analyzing this type of data. A
2x2 contingency table (Table 2-2) is set up, and the cell frequencies
one would expect by chance are statistically compared with the ones

TABLE 2-2. Contingency table of hypothetical data for 2 mantid species on cannibalism of males by females during mating.

	Female eats all or part of male:	
	Yes	No
Species A	3	7
Species B	8	2

actually obtained. But for 2x2 contingency tables when expected cell frequencies are small (less than 5), another test is preferable -- the Fisher Exact Probability Test.

For example, suppose you observe mating in 2 different species of mantid and obtain cannibalism data like that in Table 2-2. To ascertain whether the proportion of females which exhibit cannibalism is independent of species (i.e., that there is no difference in cannibalism between the species), one may use the Fisher Exact Probability Test, following the procedure given in Appendix 1.3. Note that use of this method is quite straightforward when published tables are used.

Pretend that your data were collected at the same time and under the same conditions as those in Table 2-2. Does the amount of cannibalism shown by your species of mantid differ significantly from that shown by either of the species in Table 2-2? Using the Fisher Exact Probability Test, analyze your data. (You will have 2 separate comparisons, 1 with each of those species.)

SUGGESTIONS FOR FURTHER STUDY

1. Ootheca Production

Mantids lay their eggs in foamy secretions that harden into weather-resistant oothecae. In temperate regions, the eggs overoverwinter in oothecae and hatch in spring. There is a statistically positive relationship between ootheca weight and number of nymphs emerging in T. a. sinensis (Eisenberg and Hurd, 1977).

Ootheca production and egg laying can be most readily observed if one has a group of 10 or more females. The females should be fed all that they will eat. Egg laying in T. a. sinensis may start at any time of day, and thus cannot be planned for demonstration during a particular laboratory period. Stagmomantis carolina tends to start oviposition at sunset or at night (Rau and Rau, 1913). A videotape or film of this behavior could be made for presentation during a specified laboratory period. Describe the movements of a female's cerci and abdomen during egg-laying. How are these behaviors related to the architecture of the ootheca? Speculate on how environmental variables may have affected the evolution of oothecae.

2. Eclosion (Hatching From Eggs)

Dozens to hundreds of first instar mantids emerge from each ootheca. In T. a. sinensis, first instars often emerge in the morning, usually between 0700 and 0930 hr (Eisenberg and Hurd, 1977) throughout the winter indoors and in spring outdoors. Folded-up first instars wiggle out of oothecae head first, then free themselves from a membrane "bag" that encloses each one and is attached to the bottom of the cell from which each emerges with a silken thread. In about 15 min, a mantid fully expands and walks away from its natal ootheca.

Eisenberg and Hurd (1977) found that first instars of T. a. sinensis emerge from individual oothecae in mornings during periods of as long as 14 days. Emergences involve from 21 to 392 nymphs, and last no longer than 20 min on any 1 day. Such a short emergence period may be an anti-predator adaptation.

Obtain oothecae and attempt to observe this for yourself.

SUMMARY QUESTIONS

1. Suppose you were attempting to compile an ethogram for an adult praying mantid of the species used in this exercise. What criteria would you establish in order to be able to decide when the behavioral catalog is essentially complete? At what point will the probability of observing a new and unique behavior become so low as to no longer justify the time investment required to discover it? Discuss.

2. Reconsider your grooming data in statistical terms by doing one or more of the following: a. For data on both water and 2% acetic acid, and for each sex, calculate (and present in standard format) the mean, standard error, and range. b. Pooling the water and 2% acetic acid results, compare grooming rates for males and females using the Mann-Whitney U Test. c. For the sexes combined, compare grooming rates for water versus 2% acetic acid using the Mann-Whitney U Test.

SELECTED REFERENCES

Altmann, J. 1974. Observational study of behavior: sampling methods. Behaviour 49: 227-265.

Borror, D.J., D.M. DeLong, and C.A. Triplehorn. 1981. An Introduction to the Study of Insects. 5th ed., Holt, Rinehart and Winston, New York.

Eisenberg, R.M. and L.E. Hurd. 1977. An ecological study of the emergence characteristics for egg cases of the Chinese mantid (Tenodera ardifolia sinensis Saussure). Amer. Midl. Nat. 97: 478-482.

Rau, P. and N. Rau. 1913. The biology of Stagmomantis carolina. Trans. Acad. Sci. St. Louis 22: 1-58.

Robinson, M. H. 1969a. The defensive behavior of some orthopteroid insects from Panama. Trans. R. Entomol. Soc. London 121: 281-303.

Robinson, M.H. 1969b. Defenses against visually hunting predators. Evol. Biol. 3: 225-259.

Roeder, K.D. 1935. An experimental analysis of the sexual behavior of the praying mantis (Mantis religiosa L.). Biol. Bull. 69: 203-220.

Sokal, R.R. and F.J. Rohlf. 1969. Biometry. W.H. Freeman, San Francisco.

Zack, S. 1978a. Head grooming behaviour in the praying mantis. Anim. Behav. 26: 1107-1119.

Zack, S. 1978b. Description of the behavior of praying mantis; with particular reference to grooming. Behav. Processes 3: 97-105.

Zar, J.H. 1974. Biostatistical Analysis. Prentice-Hall, Englewood Cliffs, New Jersey.

3. Sampling Methods and Interobserver Reliability: Cockroach Grooming

Robert W. Matthews
University of Georgia

A number of methods exist for obtaining quantitative samples of an animal's behavior. The decision as to which method to use in a particular situation depends on the nature of the behavior of interest, the research objective, and the resources and preferences of the investigator. If more than one person is to participate in data collection over a period of time, ease of use and interobserver reliability may influence the choice of sampling convention.

A nagging concern in behavioral research is the validity of the various sampling methods. How accurately do particular measures reflect the true values for frequency and/or duration of a behavior? Each type of sampling method has inherent biases; depending on the circumstances, each can give accurate or very inaccurate estimates of the actual case (Altmann, 1974; Dunbar, 1976; Tyler, 1979).

In this exercise we will compare the accuracy of some different sampling methods for estimating frequency and duration of cockroach grooming behavior, and assess the agreement (reliability) between different observers simultaneously sampling the behavior of the same animal. In the process, you will have an opportunity to practice different sampling methods widely used in behavior studies, and to learn to use an event recorder.

METHODS

Subject

The adult American cockroach (<u>Periplaneta</u> <u>americana</u>) is the preferred animal, though almost any species of cockroach would be an acceptable substitute. Sex is not critical, nor is the animal's past experience, rearing environment, etc.

Materials Needed

One 10-channel event recorder. For each team of 3 people: 1-5 adult American cockroaches of either sex; plastic container or small aquarium to serve as an observation arena; vaseline, Fluon, or glass or screen cover for observation chamber; stopwatch; ruler or straight-edge; cornstarch; small paper bag; pad of analysis paper; 2 colors of pencils or pens; graph paper (4 line/in grid) for part B.

Procedure

Working in groups of 3, set up an observation arena and coat the
top 2.5-5.0 cm with vaseline or Fluon to keep the roach from crawling
out. Select an adult animal from the culture at random and place it
in the paper bag with about a tablespoon of cornstarch. (A brief
"shake and bake" style agitation of the bag will insure that the
cockroach is well dusted so it will groom frequently.) Gently
transfer the roach to the observation arena.

A. Preliminary Observations and Behavior Catalog Compilation

Observe the newly dusted cockroach for 15-20 min. Take notes on
the grooming behaviors seen, and compile a catalog listing of the
grooming acts. (Consider identifying grooming acts according to the
anatomical part groomed, e.g. left antenna, right cercus). For
purposes of this exercise, it is advisable not to subdivide grooming
acts too finely; 4-8 final categories would be most workable. It is
important that all group members agree on the form and name of each
grooming act before proceeding to attempt to sample quantitatively.

B. Sampling Cockroach Grooming Behavior with One-Zero Methods

Table 3-1 lists some sampling methods appropriate for studying a
single animal's behavior. The first, ad libitum, is the method you
used for the behavior catalog compilation. The second, sequence
sampling, assumes there is a sequence; it would be a useful followup
method after this exercise is complete.
Using a sheet of 1/4" grid graph paper, construct at least 2
data sheets. Place the code numbers for the behavior acts identified
in part A in a column along the left side and write the sample
intervals horizontally along the top of the page (20 sec - 600 sec).
From the 3 one-zero sampling methods listed in Table 3-1, select
1 for the entire group to use. Using a newly dusted roach, simultaneously and independently score the occurrence (1) or non-occurrence
(0) of the grooming acts every 20 sec for 10 min. (Each person will
have recorded 30 scores for each act.)
Repeat, using another newly dusted roach and a different
one-zero method.

C. Obtaining an Ad Libitum Sample with the Event Recorder

For this part of the exercise, you will be using the event
recorder. Spend a few minutes becoming familiar with its operation.
The instructor will have previously labeled the keyboard so that each
of the 10 keys corresponds to a different behavior (up to 8 grooming
acts, resting motionless, and walking/running), and adjusted the gear
ratio so that the chart speed is about 7.5 cm/min.
Divide the keyboard among members of your team so that each is
responsible for 2 or 3 behaviors. Introduce a newly dusted roach
into the arena. After a few minute's practice with the event
recorder to coordinate your team, return the roach to its cage.
Introduce a fresh dusted roach; record the roach's grooming behavior

TABLE 3-1. Some single subject ("focal animal") sampling methods appropriate for use in this exercise.

Method	What is being sampled?	At what intervals?	How is it recorded?
ad libitum	all behavior	continuously	running log or commentary
sequence	the order in which a set of behaviors occurs	variable	whenever the sequence of interest begins, order of behaviors is recorded.
one-zero:			
a) classical time sampling	occurrence of a state and/or event through time	within any predetermined (usually short) interval; several sample periods are used in succession.	for each sample interval, score 1 if behavior occurs, 0 if it does not.
b) instantaneous ("on-the-dot")	occurrence of a behavioral state through time	at precise points in time at predetermined intervals	at precise points in time, score 1 if behavior occurs, 0 if it does not.
c) predominant activity	the extent to which an activity prevails over others	within any predetermined interval	score 1 if behavior occupies at least half of the sample interval (irrespective of distribution of its occurrence), 0 if it occupies less than half.

for 10 min. If time permits, repeat with another roach.

Before removing the section of chart used from the takeup roll, mark the beginning and end of the sample period(s) on the chart, and label each of the pen traces according to the act it records.

D. Data Analysis from Event Recorder Charts

Spread the event recorder chart out on a flat surface, and tape its corners down. First, record the actual frequency of occurrence

for each of the behaviors by counting directly from the chart record.
Then, using the ruler, draw a line of one color across the chart
every 10 sec. With the other color, simultaneously mark the chart
off at 60 sec intervals.

Construct 6 data sheets, organized something like the example
provided in Table 3-2. Then sample grooming behavior directly from
the chart record using the classical time sampling method for each
interval (10 sec and 60 sec). Thus, you will have a total of 60
samples for the 10 sec intervals, and 10 samples using the 60 sec
intervals. Repeat, using the instantaneous sampling and predominant
activity sampling techniques.

E. Reliability Indices

Each group should divide up the labor of calculating the various
indices discussed below, and together prepare a written summary of
their results and answers to the Summary Questions.

1. Percentage Agreement. A measure of interobserver reliability
which is easily calculated is the percentage agreement between all
possible combinations of observers. Using the data you gathered in
Part B and the procedure outlined in Table 3-3, calculate percentage
agreement for all combinations of observers in your team when they
used the same sample method simultaneously.

TABLE 3-2. Suggested format for analyses of event recorder data.
You will have 6 data sheets organized like this, 3 with 30 intervals
and 3 with 60 intervals.

Sampling method: Sampling interval:

Coding design: (0 = non-occurrence, 1 = occurred at least once)

Key to behaviors: 1.   6.
                     2.   7.
                     3.   8.
                     4.   9.
                     5.   10.

	BEHAVIOR SHOWN BY ROACH: SUCCESSIVE SAMPLES										
BEHAVIORAL ACTS	1	2	3	4	5	6	7	8	9	10	...30 or..60
1											
2											
3											
etc.											

---

2. Cohen's Coefficient of Agreement, Kappa. A second approach to measurement of interobserver reliability corrects for the possibility of chance agreement between observers (or between sample methods). Cohen's coefficient of agreement, Kappa, is one such measure.

$$\text{Kappa} = \frac{P_o - P_c}{1 - P_c}$$

where P_o is the observed proportion of agreement, and P_c is the chance proportion of agreement. A value of Kappa = 0.5, for example, would mean a 50% agreement between the observers after chance agreement was removed from consideration.

To calculate Kappa values, you must rearrange your data. Convert your data from part B to an agreement matrix like that shown in Table 3-4. Follow the procedures outlined in that table for each combination of observers in your team, and for each combination of sample methods. See also Aids to the Instructor for this exercise, Part A (page 247).

3. Kendall's Coefficient of Concordance, W. From ordinally ranked data (Table 3-5), one may calculate still another measure of the d degree of correlation both between different observers sampling the same behavior and between different methods for sampling the behavior. This is Kendall's coefficient of concordance, W, given by the following formula:

$$W = \frac{\Sigma(R_j - \bar{R})^2}{1/12k^2(N^3 - N)}$$

where $\bar{R}$ = mean rank = $\Sigma R_j/N$
      R_j = sum of column
      N = number of behavior categories
      k = number of observers or sampling methods

A W value of 1.0 would represent a perfect correlation. Calculate W values for observer combinations and method combinations; if additional assistance is needed in calculating W, consult Lehner (1978), pp. 134-135, and see Aids to the Instructor for this exercise, Part C (page 248).

(Note: for a measure of the correlation between only 2 sets of rankings, Kendall's tau may be used; see Appendix 1.6 for procedure.)

F. Comparative Accuracy: The Effect of Sample Interval

Use data from the event-recorded sampling period for this portion of the analysis. For each of the data sheets generated in Part D (Table 3-2), rank the behaviors as above for calculating Kendall's W. Compare your sampling data with the continuous event-recorded data by calculating 6 W values -- 2 for each of the 3 methods, one using data from the 10 sec interval samples and the

TABLE 3-3. Computation of Percentage Agreement as a measure of interobserver reliability.

A. Set up data like this (one data sheet per team of observers):

Sampling method: Total no. sample pts. or intervals:

Coding design: (0 = non-occurrence, 1 = occurred at least once)

Key to behaviors: (as in Table 3-2)

Key to observers (team members names):

      A._____   B._____   C._____

OBSERVER	BEHAVIOR	SAMPLE INTERVAL						
		1	2	3	4	5	6	....n
A	1							
	2							
	etc.							
B	1							
	2							
	etc.							
C	1							
	2							
	etc.							

B. For each time interval, score x if the observers agree as to the behavior:

OBSERVER COMBINATION	TOTAL NO. OF INTERVALS SHOWING AGGREEMENT	PERCENTAGE AGREEMENT*
A, B		
B, C		
A, C		
A, B, C		

*Percentage Agreement = Total number of observations in agreement made by a particular the combination of observers divided by the total number of sample intervals. Calculate this value for each observer combination.

TABLE 3-4. Computation of Kappa values as a measure of reliability for more than 1 observer using the same sampling method simultaneously to observe the same animal.

---

1. Convert the raw data from Table 3-3 data sheet to an <u>agreement matrix</u>:

Observer (1):

<u>BEHAVIOR</u>	<u>1</u>	<u>2</u>	<u>3</u>	<u>4</u>	Sum of entries	Proportion of total for observer 1 (P_1)
Observer (2): 1						
2						
3						
4						
sum of entries						
Proportion of total for observer 2 (P_2)						

For each cell, enter the number of sample points during which that combination of observed behaviors is recorded. If the observers are in perfect agreement, only the diagonal of the matrix will be filled in. All the off-diagonal entries represent disagreements between observers. The sum of all entries in the matrix should equal the number of sample points or intervals.

2. Calculate P_o = sum of diagonal entries divided by the total no. of entries. (This should equal the percentage agreement previously calculated for the 2 observers, serving as a check.)

3. Calculate P_c, the chance proportion of agreement. First, divide each of the summed entries from the rows and columns above by the total number of entries, to get proportion of total for each observer for each behavior. (Note that the sum of the proportions should total 1.0 for each observer.) P_c is then obtained by summing all of the products of P_1 x P_2 (that is, the products of the corresponding proportions ($P_{1,1}$ x $P_{2,1}$) + ($P_{1,2}$ x $P_{2,2}$) etc.).

4. Kappa, finally, is calculated by putting these P_o and P_c values into the formula: Kappa equals (P_o - P_c) divided by (1 - P_c). The value which is obtained is in indication of the agreement between the observers after chance agreement is removed from consideration.

5. Repeat for each combination of observers.

TABLE 3-5. Calculating Kendall's coefficient of concordance, W, for observers of a single animal using the same sampling method concurrently. Although this could be calculated for each possible combination of observers, the example below does it only for all 3 observers taken together.

1. This measure requires ordinally ranked data; that is, you need to know how common each behavior was in the eyes of each observer. Using your data sheet from Table 3-3, sum each behavior over all sample intervals for each observer.

2. For each observer, rank the behavior from 1 (equals most frequent) to n (equals least frequent).

3. Set up a table like that given below. For each behavior code, enter the rank that the particular behavior had for that observer. In the event of ties, average the ranks (that is, if 2 behaviors are tied for second rank, award each a rank of 2.5 and continue ranking the remainder starting with 4th place).

OBSERVER	BEHAVIOR CODE:									
	1	2	3	4	5	6	7	8	9	10..etc.
A										
B										
C										

Total (R_j)

4. Sum each column to get the R_j values, and average them to get $\bar{R}$. For each column, obtain $(R_j - \bar{R})^2$. Add all of these together to obtain the value for the numerator of the Kendall's coefficient.

5. For the denominator, N = no. of behavioral categories, and k = no. of observers. Calculate $1/12\ k^2(N^3 - N)$.

6. Put these values into the equation:

$$W = \frac{\Sigma(R_j - \bar{R})^2}{1/12k^2(N^3 - N)}$$

other using results from the 60 sec interval samples. Which method results in the highest W value?

Graphically compare the various sampling methods for a relatively long grooming act (e.g. antennae grooming), and again for a relatively brief grooming act (e.g. cleaning mouthpart appendages). Use actual values for the y-axis and the estimated values for the x-axis. Data points to the left of the perfect correlation line (whose slope equals 1) will represent underestimates, while those to the right are overestimates. Similarly compare a very common behavioral act with a relatively rare one.

In your write-up, indicate which sampling methods and sample intervals provide the best estimates of the "real world". Discuss possible explanations for any overestimates or underestimates obtained.

SUMMARY QUESTIONS

1. Are some sampling methods consistently better than others for producing interobserver reliability? Explain.

2. What sorts of behaviors would most accurately be sampled by each of the methods used in this exercise? Consider the effects of differing bout lengths, frequency distribution of bouts over time, frequency distribution of inter-bout intervals, and sample interval.

3. Which sampling convention appears to be the best all-around method? Why?

4. Based on your observations of cockroach grooming, generate a reasonable, researchable hypothesis tailored to each of the sampling methods used.

SELECTED REFERENCES

Altmann, J. 1974. Observational study of behavior: sampling methods. Behaviour 49: 227-267.

Cohen, J. 1960. A coefficient of agreement for nominal scales. Educ. Psychol. Meas. 20: 37-46.

Dunbar, R.I.M. 1976. Some aspects of research design and their implications in the observational study of behaviour. Behaviour 58: 78-98.

Kendall, M.G. 1948. Rank Correlating Methods. Charles Griffin and Co., London.

Lehner, P.N. 1978. Handbook of Ethological Methods. Garland STPM Press, New York.

Sackett, G.P., editor. 1978. Observing Behaviour, vol. II. Data Collection and Analysis Methods. Univ. Park Press, Baltimore.

Simpson, M.J.A. and A.E. Simpson. 1977. One-zero and scan methods for sampling behaviour. Anim. Behav. 25: 726-731.

Tyler, S. 1979. Time-sampling: a matter of convention. Anim. Behav. 27: 801-810.

SPATIAL ADJUSTMENT AND ORIENTATION

4. Orientation:
I. Thermokinesis in
Tribolium Beetles

Roger D. Akre
Washington State University

A major part of any animal's behavior is orientation. This includes both postural adjustments toward environmental space (gravity, etc.) and movements made in relation to biologically important factors (food, mate, prey, host, etc.). Not surprisingly, orientation is an active field of biological inquiry with a growing body of data, classification, theory, and terminology.

Traditionally, spatial orientation has been defined in terms of the reaction mechanism involved, i.e., the direction of locomotion in relation to the stimulus. When the locomotion is directed straight toward or away from the stimulus, the term taxis is used. When locomotion is directed at an angle to the stimulus, the response is called a transverse orientation. In kinesis, theoretically the simplest locomotory response an animal could make to a stimulus, there is no directionality to the response; however, the speed of an animal's movement and/or the frequency of turning depend on the intensity of stimulation. Thus, kineses commonly take 2 forms. Orthokinesis is variation in linear velocity (rate of movement) depending on intensity of stimulation. Klinokinesis is variation in frequency of turning depending on intensity of stimulation.

Kineses are important in the orientation of simple organisms (bacteria, Paramecium). They are helpful in analyzing and explaining the behaviors of many predators and parasites (including parasitoids) important in biological control (e.g. coccinellids, mites, syrphid larvae). Kineses are even responsible for some of the basic underlying orientation movements of insects with elaborate and complex sense organs (such as Hymenoptera, including social bees like the bumble bee).

Thermokinesis is kinetic response to differences in temperature. Given a chance, most animals will show an active preference for a certain temperature range. (It is important to remember, however, that there is probably no single optimum temperature for any animal. For example, the optimum temperature for reproduction is usually different than the optimum temperature for maximum life span.)

In this exercise, we will be observing Tribolium beetles' kinetic responses to temperature, using a water bath apparatus which maintains a temperature differential of 20°C or more across a divider of 2 mm with no gradient. We will determine frequency of turning and speed of locomotion under different temperature conditions, observe boundary reactions, and determine whether thermokinetic responses in Tribolium cause aggregations in a favorable habitat.

METHODS

Subject

Commonly called "flour beetles," Tribolium confusum Duval and Tribolium castaneum (Hbst.) are small (3-4 mm) beetles which are pests of stored products, especially grain. They are frequently found infesting cereal boxes in homes; beetles are also obtainable from commercial laboratories, biological supply companies, and the USDA. Their natural abundance and the ease with which they may be cultured make them ideal laboratory animals. Once obtained, cultures can be maintained in 1 liter jars containing whole wheat flour and closed with a cloth lid held in place by a rubber band.

A number of other insects can be used as substitutes in this experiment. However, any animal used must move slowly enough to permit tracing its path, and must not be an insect which flies readily. Small carabids, curculionids, other beetles, or even small isopods work equally as well.

Materials Needed

Beetle cultures; "used beetle containers"; plastic box, 15 x 2.5 cm (sealed) with 6 entrance tubes (Fig. 4-1); 6 pieces rubber tubing, each about 120 cm long; stop watch or watch with second hand; rapidograph pen; telethermometer; about 10 sheets of glass, 20 x 20 cm; colored pencils; 4-6 tubing clamps; multichannel thermometer with at least 2 banjo probes; tracing table or bright window; protractor; 2 immersible pumps; water heater for constant temperature bath; ice; planimeter (also called "map reader").

Procedure

Work in teams of 4-6 students. Each team should be in charge of 1 water bath apparatus. Divide responsibility among team members so that while 1 individual times a beetle, another can trace its route on 1 of the glass plates, and the rest of the group can perform the necessary measurements and calculations. Alternate so that all team members perform all roles during the course of the laboratory period.

A. The Effect of Temperature on Speed and Turning Rate

1. Adjust, with ice or hot water, both sides of the constant temperature box until 1 side is 20°C and the other is 40°C.
2. Introduce 1 flour beetle on the 20°C side, and cover the box with a glass plate. As you begin timing, trace the beetle's path on the glass, using a rapidograph pen. Indicate time along the path with a perpendicular line every 2 sec. Continue tracing the route for 20 sec, or until the beetle goes to the other side. Repeat with 4 more beetles, using a fresh piece of glass for each trial.

NOTE: Use each beetle only once, then return it to the "used beetle container". Do not include trials in which beetles travel along the edge of the test apparatus, as the edge alone determines their path. When crossing a steep gradient into a harsh stimulus, animals sometimes exhibit a "boundary reaction." If your beetle crosses from one temperature region to the other, watch for this reaction. Typically you will see the beetle make a sharp turn or sometimes even walk backwards.

3. Repeat test 5 times on the 40°C side, using the same procedure.
4. As soon as some glass plates with paths are available, begin to calculate angular velocity and distance traveled, using planimeters and protractors.
5. An alternative method of analyzing a beetle's path is to calculate the Coefficient of a Straight Line, or CSL.

$$CSL = \frac{\text{straight-line distance between start and end points of travel}}{\text{distance traveled}}$$

A CSL of 1 indicates that the beetle traveled in a straight line (no turns); the lower the CSL, the more tortuous the path.

B. Thermokinesis and Aggregation in T. confusum

1. Place 50 new flour beetles in the center of the constant temperature box. Observe their reactions and routes in a general way, as you begin timing.
2. At the end of 20 min, indicate their positions.

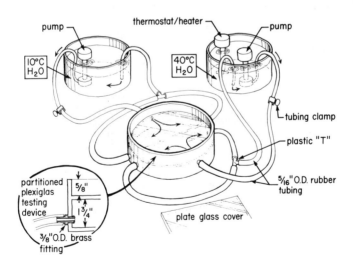

Fig. 4-1. Test apparatus used to observe thermokinesis in Tribolium beetles or other small invertebrates.

3. Repeat as time permits.

C. Effects of Extreme Cold on Thermokinesis

1. Reduce temperature of cold side of box to 10°C (or switch apparatus stations).
2. As in part A, introduce 5 flour beetles, 1 for each trial, on this side and trace their routes. Indicate each 2 sec as before. What happens if a beetle crosses the boundary to the warmer side?
3. Introduce 50 fresh beetles to the center of the box. Record their positions at the end of 20 min. How does extreme cold affect thermokinesis?

D. Thermokinesis in T. castaneum

If time permits, repeat individual trials as in part A for the red flour beetle, T. castaneum. Perform the same calculations.

E. Data Analysis

Construct graphs or tables, indicating a) speed of locomotion for the 2 species. (Use the same ordinate and abcissa at 10°C, 20°C, and 40°C); b) frequency of turning, and amount of turning in degrees (ignore sign).

SUMMARY QUESTIONS

1. Can kinesis alone explain aggregations of insects? Why or why not?

2. How efficient are kineses as compared to taxes?

3. "Adaptation to the stimulus is absolutely necessary in any animal exhibiting klinokinesis, while adaptation is undesirable in orthokinesis." Is this true? Why or why not?

4. What types of stimuli (gravity, light, heat, humidity, chemicals, etc.) are most likely to induce a kinesis in nature?

5. In this exercise, we have not controlled light or humidity. What effects might this be expected to have upon our results?

SELECTED REFERENCES

Carthy, J. D. 1958. An Introduction to the Behavior of Invertebrates. Allen and Unwin, London.
Fraenkel, G. S. and D. L. Gunn. 1961. The Orientation of Animals: Kineses, Taxes, and Compass Reactions. Dover, New York.
Gunn, D. L. 1975. The meaning of the term "klinokinesis". Anim. Behav. 23: 409-412.
Hazelbauer, G.L. (ed.). 1978. Taxis and Behavior. Elementary Sensory Systems in Biology. Chapman and Hall, London.
Jander, R. 1963. Insect orientation. Annu. Rev. Entomol. 8: 95-114.

5. Orientation:
II. Optomotor Responses
of Dragonfly Naiads

Roger D. Akre
Washington State University

While riding in a car, we tend to follow passing scenery with our eyes. This is an involuntary tendency, and can be controlled by consciously fixing our eyes in 1 position. However, if this is done while looking out the side window of a rapidly moving vehicle, the sensation of objects passing rapidly over the visual screen is unpleasant and frequently leads to motion sickness. This involuntary movement of the eyes to follow objects, the optomotor reaction, also occurs in insects. However, since they cannot move their eyes, they must move their entire body to follow the moving object.

The visual fields of the compound eyes of Aeschna larvae (and probably of all dragonfly naiads) overlap so binocular vision occurs, and with this type of vision comes an ability to judge distance. When potential prey enters the visual field of an immature dragonfly, the naiad will usually turn directly toward the prey and approach it slowly until certain ommatidia (fixation ommatidia) in both eyes are stimulated. The optical axes of these ommatidia intersect at a particular distance from the eye, the exact distance the naiad can extend its labium to grasp prey. Similar measurements of distance are used by other predators such as Notonecta, salticid spiders, and praying mantids.

An optomotor apparatus (Figs. 5-1, 5-2) provides an experimental way to move "landmarks". In this exercise, we will use 2 types of optomotor apparatus to investigate the optomotor response of dragonfly naiads. If time permits, we will also investigate the strike response of naiads in capturing prey.

METHODS

Subject

Dragonfly naiads can be found in still water (permanent ponds, lakes) and in fast moving streams. While naiads collected from streams will exhibit a very strong optomotor response, these insects are difficult to maintain for more than a few hours without moving water. Therefore, it is best to use the hardier still water forms.

All dragonfly naiads are predators, and naiads of the larger species will attack small fish or each other, if given the chance. They should not be kept in small containers together.

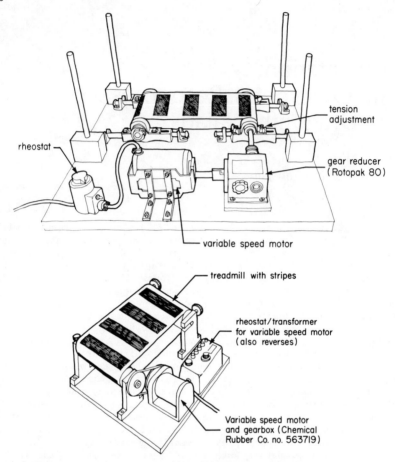

Fig. 5-1. Alternative designs for a treadmill optomotor apparatus particularly suited to study the effects of pattern position and size.

Materials Needed

6 dragonfly naiads; treadmill optomotor apparatus (Fig. 5-1); patterned paper for treadmill, with black stripes of 6 mm, 12 mm; 15 cm diam. plastic dish with vertical stripes, hooked to variable speed drill (Fig. 5-2); 6 small plastic dishes to hold naiads. Optional: wire, beeswax, heat source; artificial prey; minnows, guppies, or aquatic arthropods of various sizes.

Procedure

Students are divided into 2 teams. One team will run the experiment using the treadmill, the other the optomotor apparatus made with the drill.

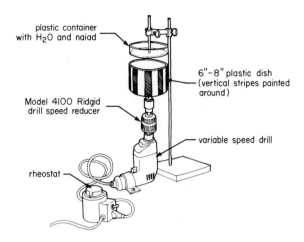

plastic container
with H₂O and naiad

6"-8" plastic dish
(vertical stripes painted
around)

Model 4100 Ridgid
drill speed reducer

variable speed drill

rheostat

Fig. 5-2. Optomotor apparatus using a variable speed drill as its motor. This design is suitable for testing effects of pattern speed upon the optomotor response of aquatic invertebrates.

A. The effect of pattern position and size (Team 1)

Place the treadmill in a horizontal position. Suspend a clear plastic container containing a naiad 5 cm above the treadmill pattern (6 mm pattern). Start the treadmill (slowly!). Does the naiad try to follow the stripes? If not, suspend the naiad container 2.5 cm above the treadmill pattern. What effect does varying the speed have on the naiad's perception of the moving pattern? Change the pattern to the 12 mm stripes and repeat.

After testing the naiads with the treadmill pattern moving below them, suspend the naiad container 2.5-5.0 cm away from the treadmill, tipped onto its side so the visual pattern is at the same level (rather than underneath) the naiad. Does the naiad respond differently? Repeat with 12 mm stripes.

B. The effect of pattern speed (Team 2)

Test a naiad for optomotor response by suspending its container inside the patterned container hooked to the drill. Start the pattern slowly revolving. Observe the response of the naiad. Slowly increase the speed of revolution of the pattern until the naiad no longer responds.

(Optional:) Sometimes the reaction of the naiad to the moving pattern is more easily observed by gluing the naiad to a solid support (such as wire, using beeswax), and holding the naiad stationary within its container of water. If this is done, make sure the legs of the naiad are still able to move freely.

SUGGESTIONS FOR FURTHER STUDY

To investigate the strike response of dragonfly naiads to prey, each student should obtain a naiad in a 15 cm diameter plastic dish (with cover) containing water. The laboratory instructor has prepared a series of artificial prey items. Using these, test the strike response of your naiad. Live prey (minnows or guppies, aquatic arthropods of various sizes) may also be furnished by the instructor.
Investigations might include:
1) Size threshold -- how large can the prey be? How small?
2) Must the prey be moving? What happens when the prey is still, but a pattern is moved behind the prey?
3) Sequence of attack -- how far away is the prey when the naiad detects it? Then what happens? (Describe.)
4) Distance to prey when attacked -- does this distance vary?

SUMMARY QUESTIONS

1. What is one of the main functions of the optomotor response in insects living in fast flowing streams?

2. Might the optomotor response be used to determine whether insects detect color? How?

3. What function does the optomotor response perform in flying insects such as locusts?

4. Mantids in jars also show optomotor-like responses when crickets walk along the insides of their jars. What additional roles does this suggest for the adaptive advantages of this response? Are these proximal or ultimate causation hypotheses?

SELECTED REFERENCES

Carthy, J.D. 1958. An Introduction to the Behavior of Invertebrates. Allen and Unwin, London.
Lillywhite, P.G. 1980. The insect's compound eye. TINS (Trends in Neuro Science). pp. 169-173.
Mazokhin-Porshnyakov, G.A. 1969. Insect Vision. Plenum, New York.
Pritchard, G. 1965. Prey capture by dragonfly larvae (Odonata: Anisoptera). Can. J. Zool. 43: 271-289.
Pritchard, G. 1966. On the morphology of the compound eyes of dragonflies (Odonata: Anisoptera), with special reference to their role in prey capture. Proc. Roy. Entomol. Soc. London (A) 41: 1-8.
Snodgrass, R.E. 1954. The dragonfly larva. Smithson. Misc. Coll. 123(2): 1-38.

6. Patterns of Spatial Distribution in Field Crickets

William H. Cade
Brock University, Ontario

Patterns of animal distribution in space may generally be classed as random, clumped, or uniform. In a <u>random</u> pattern of distribution, each unit of space has an equal probability of having an organism present. In a <u>clumped</u> pattern, the presence of 1 individual greatly increases the probability that another individual is nearby. Under a <u>uniform</u> pattern of spatial distribution, organisms occur at regular intervals from one another.

Spatial distributions may indicate the influence of abiotic factors and other ecologically important phenomena which have but a minor behavioral component. In many animals, however, this is not the case. Patterns of animal spatial distribution are often indicative of underlying behavioral responses. For example, a high degree of clumping may result from individuals being attracted to signals produced by members of their species. Thus a particular forest tree may have a much greater density of beetles already present on the host tree. At the other extreme, male birds may be uniformly distributed throughout the forest if they are repelled by signals and aggressive behavior typical of territoriality. In these cases, determination of the spacing pattern is often the first step in analyzing important behavioral characteristics of a species.

The major objectives of this exercise are to gain an understanding of the nearest-neighbor and quadrat techniques, 2 methods commonly used to determine spacing patterns. It is important to use both methods since they have different advantages and limitations. This exercise could be used with practically any group of organisms, but is specifically intended for acoustical insects. Field crickets are ideal since they are usually quite common in most areas during spring, summer, and early fall, and since they readily indicate their position in space by producing nearly-continuous acoustical signals. In addition, a large amount of information is available on the behavior of crickets and their relatives, and important inferences can be drawn from the patterns observed in this exercise.

METHODS

Subject

Field crickets are widely distributed over North America. In the northeastern United States and southeastern Canada, the field cricket, <u>Gryllus</u> <u>pennsylvanicus</u> Burmeister (Orthoptera: Gryllidae), is a good choice. Male <u>G</u>. <u>pennsylvanicus</u> produce the rhythmic chirping sound which functions to attract females. Males call at night, but switch to daytime signalling if the night temperature is below 10-12°C. Before the lab meets, a suitable population of crickets should be located and peak calling time determined.

Materials Needed

Surveyor's tape; pieces of string 100 m long, marked at 10 m intervals; tape measures; flashlights if exercise is performed at night. Optional but helpful: a statistics textbook such as Schefler (1969).

Procedure

A. Practice Listening to and Locating Crickets

With the help of another student, practice locating individual calling males. Both investigators first stand 3 to 4 m apart, and about 10 m from where the cricket is initially believed to be. Slowly and very softly walk in the direction of the cricket. If the exercise is performed at night, shine your flashlight to the point where you hear the cricket. Each person thus takes periodic "fixes" on the cricket. Since crickets are somewhat ventriloquial, the male will usually be located where the line of sight of the 2 investigators crosses. With practice, however, a person working alone can approach within 1 m of a calling cricket. If the cricket stops calling before its position is located, move on to another cricket and return later to the silenced one. Remember, walk softly.

B. Marking the Location of Crickets

Before the laboratory session, a 100 m x 100 m portion of the field will be flagged. Working in pairs, mark the location of all calling crickets in the field. Follow the same procedure practiced earlier, but now as a marker tie a piece of surveyor's tape to a piece of vegetation next to the cricket. Assign each cricket a number, and write it on his marker. The marked locations will now be used in collecting the following data.

C. The Grid System

One popular method of determining spacial patterns uses a grid system in which the field is mapped in a series of quadrats. Set up the grid by stretching a 100 m piece of string from 1 corner of the measured 100 m square to an adjacent corner. The string is marked at

10 m intervals. Place a flag at each of these intervals. Each side of the 100 m square will be marked in this way. Once the outside measurements have been made, mark the inside coordinates in a similar way. The 100 m square is now divided into a series of one hundred 10 m x 10 m quadrats.

Draw a map of the grid in your notebook. Walk through the quadrats and mark the position of each cricket on your map. It is important to map these positions as accurately as possible. That is, if the cricket is in the lower right portion of a quadrat, this should be represented on your map. These data will later be used in calculations involving the Poisson distribution.

Remove all flags used to mark the quadrats, but leave the cricket markers and the 4 flags marking the corners of the field.

D. The Nearest-Neighbor Technique

A second popular method of determining spacing patterns uses the distance between nearest neighbors. Use a tape measure to determine these distances. Begin in 1 corner of the field and make sure that each male has 1 measurement. In keeping with standard practice and for use in the later calculations, either take your measurements in meters or convert them from feet to meters. To avoid duplications, write the number of each cricket in your notebook and the distance he is removed from his nearest neighbor. In most cases, which neighbor is the nearest can be judged visually, but sometimes you may need to take measurements between a male and several others to determine who is the nearest. Males widely separated from other males will have a measurement to a nearest neighbor, who in turn will be closer to another male. But in many cases, 2 males will be each other's nearest neighbor. The important point is that each male will have 1 measurement.

All flags placed in the field can now be removed, and the collected data used in the following computations.

CALCULATIONS

1. Grid Data

a. Set up a table with columns headed in the following way:

x	f	fx	P(x)	F	F-f	$(F-f)^2/F$	x-m	$(x-m)^2$	$f(x-m)^2$
0									
1									
2									
3									
etc.									

b. The f column gives the frequency of occurrence of quadrats with x males. Count the number of quadrats containing 0 males and enter this number into the appropriate place in column f. Then count the number containing 1 male, 2 males, and so on.

c. Calculate the mean number of males per quadrat (m). To do this, multiply each term in the x column by the corresponding term in the f column, and enter each product in the fx column. Sum all the products in the f column and divide the resulting number of males by 100 (the total number of quadrats); this gives you m.

d. Calculate the frequency of males per quadrat expected if the population is randomly distributed. First, generate the mathematical approximation of a random distribution for the cricket population by calculating for each x the probability that any 1 quadrat contains that number of males, P(x). This probability can be found by applying the following formula:

$$P(x) = e^{-m}(m^x/x!)$$

where m = mean no./quadrat; x = no. in quadrat; e = 2.71828, the base of natural logarithms. (Look up e^{-m} in a table of exponential functions.)

For example, for a mapped population of crickets with an average of 0.78 males/quadrat, the probability of a quadrat chosen at random having 3 males is

$$P(3) = 2.71828^{-0.78} \times 0.78^3/3! = 0.458406 \times 0.474552/6 = 0.036.$$

Therefore, the probability of a quadrat having 3 males is relatively low. By comparison, the probability of a quadrat having 1 male is 0.218, and 0 males is 0.458.

Each P(x) you calculated is now entered in the P(x) column in your table. Use these probability terms to calculate the expected frequency distribution (F) if the males are randomly distributed. To do this, multiply each P(x) value by the total number of quadrats (100), and enter the products in the F column.

e. Inspect the F column of the table and lump together adjacent terms until they total at least 2.0. When terms in column F are lumped, the corresponding terms in other columns must also be lumped. For example:

This:

x	f	F
4	0	6
5	1 lumped	1.3 lumped
6	3	0.9

Gives you this:

f	F
0	6
4	2.2

f. To compare the observed frequency distribution with the distribution expected from chance (see chi square goodness of fit, Appendix A), subtract each term in the f column from the corresponding term in the F column. Square the difference and divide by F. Your table will now have filled in the 2 columns headed F-f and (F-f)²/F. To determine whether these differences are statistically significant, perform a chi square test on the data. The degrees of freedom are the number of terms in the (F-f)²/F column minus 1. Sum the terms in this column to get the value of chi square, and look up the corresponding p value in an appropriate table. For this exercise, use a 5% level of significance. Make a probability statement regarding the comparison of your data with what is expected under random conditions.

g. Calculate the variance (s^2) of your samples to determine how much your data vary from the mean from quadrat to quadrat. Use the following formula:

$$s^2 = \frac{\Sigma f(x-m)^2}{N-1}$$

If the population is dispersed randomly, s^2 will approximately equal the mean, and the chi square probability should have been greater than 5%. If the population is uniformly disperse, the variance will be smaller than the mean. A population having a clumped dispersion pattern will have a variance larger than the mean.

h. Check that you have everything. For this portion of the exercise, you should have a map of the grid showing the location of all crickets, a table of values, the sample mean, the sample variance, a statement of whether or not the observed frequency distribution fits the expected distribution if the population is randomly distributed, and a statement regarding the type of distribution determined from these calculations.

i. One shortcoming of the grid technique is that the choice of grid size usually affects the type of distribution. To demonstrate this effect, perform the same calculations but with a grid size of 5 m², half the size used in the earlier calculations. That is, the total number of quadrats will now be 400. This calculation can be performed by simply dividing each quadrat on your map into quarters, since the positions of all males were marked accurately. You may also wish to determine the effects of doubling the grid size to 20 m². For this portion of the exercise, you should have the same information reported above for the 10 m² grids.

2. The Nearest Neighbor Data

The grid technique relied on a comparison between the distribution of crickets expected by chance (that is, if the population is randomly distributed) and the observed distribution. The nearest neighbor analysis makes the same general comparison, but uses intermale distances rather than the frequency of males per quadrat. This is the real advantage of using the nearest neighbor technique.

a. Review your raw data, which consists of a distance measurement for each male in the population sampled. Each observed distance measurement is termed r_o. Calculate the mean r_o for the population. That is, determine

$$\bar{r}_o = \frac{\Sigma r_o}{N}$$

where N is the number of males sampled, and Σr_o is the sum of all intermale distances in the sample population.)

b. Determine the density (d) of males in the population by dividing the total number of males by the area sampled (100 m²). This measurement is the average number of males per m² in the population.

c. Determine the average distance between nearest males to be expected if the population is distributed randomly. The average expected distance is determined by the following formula:

$$\bar{r}_e = \frac{1}{2\sqrt{d}}$$

(For derivation of this formula, see Clark and Evans, 1954.)

d. The value R is the measure of the degree to which the sample population departs from randomness. If the population is randomly distributed, R will be close to 1. In a clumped population, R is smaller than 1, whereas in a uniformly dispersed population R is larger than 1. Determine the ratio

$$R = \frac{\bar{r}_o}{\bar{r}_e}$$

e. Test the statistical significance of any departure from randomness by computing the standard score (z):

$$z = \frac{\bar{r}_o - \bar{r}_e}{\text{standard error of } \bar{r}_e}$$

The standard error of $\bar{r}_e$ is given by $\dfrac{0.26136}{\sqrt{Nd}}$

Use the z value to determine the probability that a greater difference between the observed and expected intermale distances will result from sampling errors. The 5% level of significance is again used. This level of significance is reached when $z = 1.96$. Therefore, the calculated difference between $\bar{r}_o$ and $\bar{r}_e$ is judged to be statistically significant when z is 1.96 or greater. (The precise probability attached to your calculated z values can be determined by consulting a z table.)

f. Check that you have everything. For this portion of the exercise, you should have a table showing the nearest neighbor distances for each male in the sample population, a computation of the nearest neighbor statistic, a decision regarding the type of distribution determined in this analysis, and a probability value attached to this decision.

SUMMARY QUESTIONS

1. How does quadrat size affect the results obtained from the grid system of determing space patterns? What factors should be taken into consideration in choosing a particular grid size?

2. Is the decision of the nearest neighbor technique affected by the total area sampled? What factors should be taken into consideration in choosing the sample area?

3. What are some of the advantages and disadvantages of both measures of dispersion? Which technique is best suited for use with field crickets? What attributes of other organisms might lead you to use the other measure of dispersion? If you reached different conclusions in the nearest neighbor and grid analyses, why might this have happened? Which conclusion was probably the more accurate?

4. In addition to the inherent problems of both techniques, what are some other sources of error in studying field cricket spacing patterns? In what kinds of organisms might these problems be minimized?

5. What kinds of behavior might result in the distribution pattern observed in this exercise? (This is an important question; the selected references or a general book on animal behavior should be helpful in formulating this speculation.) What are some ways to test or demonstrate the importance of field cricket behavior in producing the resulting distribution pattern?

SELECTED REFERENCES

Alexander, R.D. 1961. Aggressiveness, territoriality, and sexual be- havior in field crickets (Orthoptera: Gryllidae). Behaviour 17: 130-223.
Alexander, R.D. 1975. Natural selection and specialized chorusing behavior in acoustical insects. In D. Pimentel (ed.). Insects, Science, and Society. Academic Press, New York, pp. 35-77.
Brown, J.L. and G. H. Orians. 1970. Spacing patterns in mobile ani- mals. Annu. Rev. Ecol. Syst. 1: 239-262.

Cade, W. 1981. Field cricket spacing, and the phonotaxis of crickets and parasitoid flies to clumped and isolated cricket songs. Z. Tierpsychol. 55: 365-375.

Campbell, D.J. and D.J. Clarke. 1971. Nearest neighbour tests of significance for non-randomness in the spatial distribution of singing crickets (Teleogryllus commodus (Walker)). Anim. Behav. 19: 750-756.

Clark, W.C. and F.C. Evans. 1954. Distance to nearest neighbour as a measure of spatial relationships in populations. Ecology 35: 445-453.

Morris, G.K. 1971. Aggression in male conocephaline grasshoppers (Tettigoniidae). Anim. Behav. 19: 132-137.

Morris, G.K., G.E. Kerr, and J.H. Fullard. 1978. Phonotactic preferences of female meadow katydids (Orthoptera: Tettigoniidae: Tettigoniidae: Conocephalus nigropleurum). Can. J. Zool. 56: 1479-1487.

Otte, D. 1977. Communication in Orthoptera. In T.A. Sebeok (ed.), How Animals Communicate. Indiana University Press, Bloomington. pp. 334-361.

Otte, D. and A. Joern. 1975. Insect territoriality and its evolution: population studies of desert grasshoppers on creosote bushes. J. Anim. Ecol. 44: 29-54.

Pielou, E.C. 1960. A single mechanism to account for regular, random and aggregated populations. J. Ecol. 48: 575-584.

Schefler, W.C. 1969. Statistics for the Biological Sciences. Addison-Wesley Publ. Co., Reading, Mass.

Southwood, T.R.E. 1978. Ecological Methods. With Particular Reference to the Study of Insect Populations. 2nd ed. Halstead Press, John Wiley and Sons, New York.

7. Object Orientation in Mealworm Beetles

Rudolf Jander
University of Kansas

Virtually all mobile organisms are capable of approaching objects that are resources to them. The execution of this capacity is called object orientation. If generality were used as the criterion for importance, object orientation could be considered to be truly the most important of all behaviors.

Object orientation has 3 major components: ranging, local search, and approach. The latter implies fairly accurate knowledge about the position of the object, and is not investigated in this exercise. Search as a general term includes the entire alternative class of behaviors resorted to whenever localization is ruled out. Ranging is the type of search seen when the organism is almost or completely without knowledge about where to find a resource. During ranging, ground coverage is maximized. (How might this best be achieved? How, therefore, can ranging be recognized?)

Local search replaces ranging whenever the searcher encounters cues to the proximity of some resource in want. During local search (also called "area restricted search"), a restricted area is more thoroughly investigated. The pattern of locomotion which results is strikingly different from ranging behavior. (Can you imagine a pattern of locomotion that best serves local search?) Typical local search is one of the most randomly (unpredictably) structured of all behaviors, as we will learn by experience during this exercise.

METHODS

Subject

Virtually any species of insect deprived of some resource could be used for the following experiments. However, adult mealworm beetles (Tenebrio molitor) offer many advantages. They rarely fly away, they readily walk at room temperature, they are little disturbed by a lively group of human observers, and they are readily and inexpensively available everywhere year-round.

Materials Needed

Mealworm beetle culture; pieces of cardboard, 50 cm x 50 cm; filter paper cut into small (50 mm²) equal pieces ("confetti"); camel's hair brushes; beaker with water; glass rod or pipette; elevated T-mazes and L-T-mazes (for construction hints, see Aids to the Instructor). Optional: turntable or Lazy Susan; statistics textbook.

Procedure

A. Observations of Ranging and Local Search

In order to observe the transition from ranging to local search, you need to have the beetle walk some distance on a plain cardboard surface initially, and thereafter encounter a moist (not wet) piece of filter paper. This stimulus is then "interpreted" by the beetle as a cue for resource proximity.

In order to increase the chances for encountering such a cue, it is expedient to release the beetle at the center of a circle of such cues. A practical arrangement is a circle of 10 cm radius, with pieces of filter paper with every 3 cm along the periphery (Fig. 7-1). Construct such a stimulus ring.

Release 1 beetle at a time at the center of the circle. It will proceed in a more or less straight line (ranging) until it hits a piece of moist filter paper, the cue. This event initiates a convoluted local search path which increases the chance of finding other nearby sources of moisture. If the beetle happens to exit between 2 pieces of filter paper, replace it into the center. Repeat the whole procedure with more beetles until the "rules" of the beetles' behavior have been learned by "observational induction."

NOTE: If contacting a piece of moist filter paper fails to trigger local search behavior, the explanation is usually not the lack of proper motivation (thirst), but competing escape motivation, most likely induced by overly bright light or by mishandling.

Two measures are likely to improve your success. Since mealworm beetles prefer darkness, they are more prone to go into local search behavior if the light is dim. Furthermore, beetles should be handled as gently as possible. Do not squeeze them between fingers or forceps to transport them. Instead, place a strip of cardboard in front of a beetle, then prod the beetle with a paint brush from behind until it crawls onto the cardboard. Use the cardboard as a vehicle for replacement and displacement.

B. Analysis of Local Search Behavior

Once the conspicuous difference between the straight ranging path and the convoluted local search path is firmly engrained into your mind, as a group discuss possible adaptive functions of these behaviors in natural environments. Then the question of the underlying behavioral mechanisms should be raised. The structure of

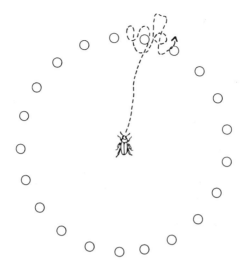

Fig. 7-1. Circular arrangement of moist pieces of filter paper into the center of which a mealworm beetle has been placed. The dashed line is a typical path of a beetle so released.

convoluted search paths will be analyzed more quantitatively during later experiments with foraging honey bees (Exercise 12). With such data at hand, stochastic modeling (Exercise 32) is also profitable.

C. Ranging Behavior

Ranging is more than just forward walking. Specific control mechanisms are involved that ensure staying on course.

Force a mealworm beetle to make a turn on a L-T maze (Fig. 7-2). After the forced turn, the beetle has a choice at the T-intersection: to turn either left or right. Will the beetle remember the forced turn and counterturn, thus trying to keep the overall course straight? It frequently will, but at other times will not. The experiment must be repeated at least 30 times, using a different beetle each time, and then statistically analyzed. An appropriate statistical test is the chi square one-sample goodness-of-fit test; its use is explained in Appendix 1-2.

D. Sources of Systematic Error

Statistical analysis is not enough to draw a safe conclusion. It takes care of the <u>random errors</u> only, not possible <u>systematic errors</u> with respect to the question asked. If the beetles, for instance, have been forced to turn left and then turn preferentially to the right at the T-intersection, this might be due to some side tendency (laterality or "handedness"; see Exercise 30) of the beetles. What is the best control against this possible error? Simply set up another maze of mirror image configuration (Fig. 7-2), and test all beetles with both mazes.

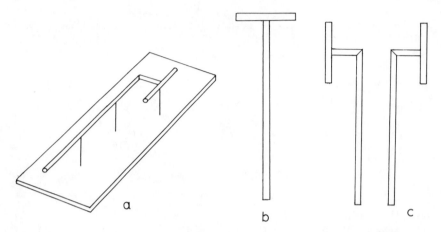

Fig. 7-2. Elevated mazes to analyze ranging in mealworm beetles. The support of such mazes on needles is illustrated in (a). The shapes of T-mazes and L-T-mazes are sketched in (b) and (c), respectively.

Are we now safe with our conclusion? Not yet. Think for a moment and see if you can find out for yourself before reading on.

Suppose the mazes had been placed in such a way that the beetles started away from the experimenter. It could have been that they somehow sense him, and escape from him as a potential predator. (Any other stimulus in the environment might have an equivalent repellent or attractant effect.)

Shielding the beetle from all conceivable stimuli is difficult, if not almost impossible. Is there a simple solution to this experimental dilemma? Yes, just turn the mazes 180 degrees and test whether or not the beetles reverse their turning direction at the T-intersection. Is the turning sense maze-determined or environment-determined?

SUGGESTIONS FOR FURTHER STUDY

Once counterturning has been discovered, many questions about its structure and quantitative properties can be asked. This is a fine opportunity for independent initiative on the part of students. Only one of many follow-up questions is described here.

As an animal moves, the stimuli it perceives are of 2 types, those produced by its own movement (reafferent) and those produced by movement in the external world (exafferent). Thus, if a beetle is forced to turn, 2 sets of events occur simultaneously which can be separated experimentally. Suppose, for instance, that the beetle on the maze turns to the right. In addition to its body turning right, the beetle's internal image of its environment turns to the left. Based on this, we ask: How is the beetle detecting and remembering its deviation from a course, indirectly by some reafference or directly by the execution of the turn on the maze?

It is easy to control light as a reafferent stimulus, for example. It can be turned off while the beetle executes the forced turn. Does this eliminate counterturning? Conversely, the environment can be made to turn, and thus cause an afference equivalent to the previous reafference. Place the straight maze on a turntable, and while the beetle is crawling down the straight section, turn the whole maze with the beetle on it 90 degrees, causing the environment to counterrotate the same amount relative to the sense organs of the beetle. Is this manipulation followed by counterturning at the T-intersection?

SUMMARY QUESTIONS

1. Explain the concepts of "ranging", "local search", and "approach". Discuss the biological significance of the orientational behaviors so defined.

2. Objects serving as resources -- such as flowers for bees -- may be distributed in the plane in 1 of 3 ways: randomly, regularly (all distances between adjacent objects are equal), or clumped (clustered). For which of these modes of dispersion is ranging without local search a sufficient search strategy? Which mode of dispersion calls for a search strategy including both ranging and local search?

3. There are several different orientational methods with which an insect may maintain a straight course while walking or flying. What are some of those methods?

4. For what reasons do we call the local search of the mealworm beetle a "random search"? Use paper, a pencil, and your imagination to create some highly systematic (non-random) local search paths.

SELECTED REFERENCES

Jander, R. 1975. Ecological aspects of animal orientation. Annu. Rev. Ecol. Syst. 6: 171-188.
Krebs, J.R. 1979. Foraging strategies and their social significance. In P. Marler and J.G. Vandenbergh (eds.), Handbook of Behavioral Neurobiology. Plenum Press, New York, pp. 225-270.
Lehner, P. N. 1979. Handbook of Ethological Methods. Garland STPM Press, New York.
Meyer, M. E. 1976. A Statistical Analysis of Behavior. Wadsworth Publ. Co., Belmont, Calif.
Pyke, G., H. R. Pulliam, and E. L. Charnov. 1977. Optimal foraging: a selected review of theory and tests. Quart. Rev. Biol. 52: 137-154.
Wilson, D.M. and R.R. Hoy. 1968. Optomotor reaction, locomotor bias, and reactive inhibition in the milkweed bug Oncopeltus and the beetle Zophobas. Z. vgl. Physiol. 58: 136-152.

8. The Locomotion of Grasshoppers

Robert Franklin
University of Oregon

Animals have evolved a variety of body structures and associated behaviors with which they can propel themselves through the media of their habitats. We can recognize 2 general sources of information used to guide their movements:

1) Information from the environment can be used as a frame of reference for whole body movements. These behaviors are called orientation.

2) Animals can use their own body axes as a frame of reference for the movements of their body parts (typically their limbs). These movements are called coordination.

Locomotion is accomplished by the coordinated activity of limbs (or body parts). Coordinated means they act together in a smooth and concerted way. The necessity of coordination to attain purposeful movement can be quickly understood if one imagines trying to swim from 1 end of a pool to another by simply thrashing arms and legs about in all directions; traversing the pool under these conditions is not impossible but is highly unlikely. Directed, purposeful movements of the body require that animals constrain the activities of their limbs to provide propulsion in the desired direction. You will study these constraints and their effects by observing and experimenting with the locomotion of grasshoppers. Grasshoppers have adapted their limbs and behavior for locomotion in the 3 principal media which animals encounter in nature: land, water, and air. They can walk, jump, swim, and fly. You will study the first 3 of these locomotory activities in some detail.

METHODS

Subject

Any of a variety of grasshoppers (sometimes referred to as locusts, a name commonly applied to the genus Schistocerca) may be used. Melanoplus, the largest genus of spur-throated grasshoppers (which includes our most common grasshoppers and the ones that are the most destructive to crops), is suggested.

Grasshoppers may be either collected or ordered from a biological supply house, and maintained as outlined in the Aids to the Instructor. You will need 1 or 2 healthy young adult grasshoppers per student.

Materials Needed

150 watt spotlight; broomstick or wooden dowel, approx. 1 m long; dental wax (such as Surgident Periphery Wax); dissecting scissors; optomotor apparatus; porcelain bowl, approx. 60-75 cm in diameter; white poster board, black construction paper, and rubber cement for constructing optomotor and swimming orientation patterns; acrylic preservative for patterns, if desired.

Procedure

Before studying locomotion by these animals, you should know something about the ways in which their legs can move. Since insects are bilaterally symmetrical, examining the 3 legs on one side tells you how legs on both sides can move. Locate the grasshopper legs' 4 major joints (Fig. 8-1): coxa/body, trochanter/coxa, femur/tibia, and tibia/tarsus (the trochanter/femur is fused in grasshoppers). The coxa can move in 3 planes with respect to the body: up and down (dorsoventrally), back and forth (rostrocaudally), and by rotation. This joint is referred to as having 3 degrees of freedom. The other 3 joints move only by flexion and extension in a plane. For example, during flexion the hind tibia is drawn up against the ventral aspect of the femur on which it articulates; during extension, the tibia swings away from the femur. Therefore, each of these joints has 1 degree of freedom.

Combined, there are 64 different ways in which the joints of 1 leg can move relative to another. To get an idea of the different ways in which its legs can move, pick a grasshopper up by grasping it at the thorax with your thumb and forefinger. As the animal struggles to escape, notice the ways in which its legs are used. In particular, observe the great flexibility of movement of the hind legs. These legs are greatly enlarged relative to the other 4. The enlarged femora accommodate tibial extensor muscles which can develop up to 1 kg of thrust in some species! These combined with the elongated tibiae allow the animal to jump great distances.

A. Handling the Animals

Grasshoppers' ability to jump -- an effective behavior in response to adverse stimuli -- makes the success of the following exercises dependent on how you handle your animals as you experiment with and observe them. When placed in a novel situation (such as a table top for observation) and "threatened" by your approaching hand or rapid movements nearby, grasshoppers will jump. Unless you wish to study escape responses, these jumps can interfere with your work. They can be minimized by following these suggestions:

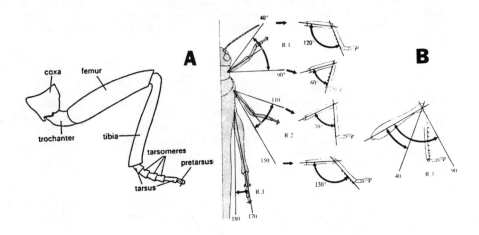

Fig. 8-1. A. A grasshopper leg showing the segments and joints. (From Chapman, 1971) B. The three legs on one side of a grasshopper. The angles shown are the ranges of movement of these legs during normal straight walking forward. (From Burns, 1973)

1) Move slowly around animals you are observing. Be sparing in handling them.

2) Darken the room in which you are working except for the immediate working area. This limits the visibility of the grasshopper to objects in its immediate area, thus hiding observers and their movements.

3) When possible, place the animal you wish to work with into the experimental situation a few minutes before actual experiments and observations begin. This allows it to adapt to local objects and conditions.

4) If your grasshopper persists in jumping despite your care, you can tether it with string and allow it to "tire" of jumping. This is not usually desirable, since tired animals do not perform as well as fresh ones. An alternative is to interfere with the ability of the hind legs to flex fully and develop the muscle tension for a jump. You can do this by placing a small piece of wax on the ventral surface of the hind femora just anterior to their articulation with the tibiae. This should not interfere with the animal's ability to walk, and it will become accustomed to the wax if allowed a few minutes. In general, careful handling, slow movements around the animals, and patience will reward you with successful experiments.

B. Walking

There are 3 basic kinds of walking which grasshoppers perform: straight walking (forward and backward), lateral walking (sideways and in either direction), and rotation (turning in place). All these can intergrade. For example, straight walking forward and rotation can be combined to result in a curved path of body movement. In the laboratory, however, it will be most instructive if you attempt to isolate each type as much as possible.

Forward Walking. Straight forward walking can be evoked by placing your animal on a smooth, flat surface and, should it not walk spontaneously, moving the spotlight close to its "tail". (Be careful not to actually burn the animal!) It will respond to this aversive stimulus by walking away. As it does so, see if you can discern which legs step together.

Insects generally step with 3 legs at a time, arranged in a tripod in which the front and rear legs of 1 side and the middle leg of the other move in synchrony. The legs of each thoracic segment alternate in stepping, and the net effect is 2 alternate tripods. Each of the legs performs the same basic activity to propel the animal: each is lifted in the air and swung in the direction in which the body is traveling. (This is called the "return stroke" of the leg.) Each is then placed on the ground, and moves backward with respect to the direction of travel, propelling the animal. This is the "power stroke" of the leg. These 2 types of strokes occur in succession to form cycles of locomotory behavior for each leg (Fig. 8-2).

The hind legs of grasshoppers, being quite a bit larger than the other 4, need not step as often to cover the same amount of ground. A ratio of 2 steps for the front and middle legs to 1 for the hind legs is common. When all 6 legs are stepping, the tripod pattern is used. When only the front 4 step, a quadrupedal pattern appears; this is the same basic pattern which terrestrial vertebrates use for locomotion. (In fact, grasshoppers can use this pattern exclusively, as you can verify by amputating the hind tibiae just below their articulations with the femora.) The ability of behaviors to change to meet changing environmental conditions is demonstrated by the grasshoppers' ability to walk with either 6 or 4 legs and to alter the coordinated stepping of its legs appropriately.

Backward and Rotational Walking. Place your animal on a broomstick or wooden dowel which is horizontal and supported at each end a few inches above the ground. Place the spotlight near the head of the animal and in front of it. In response, the animal will do 1 of 2 things: 1) it will walk backwards away from the light and heat source, or 2) it will turn on the spot, and walk the other way. The latter behavior is rotational locomotion which, at the instructor's discretion, can be investigated more fully as a separate topic. You may have to try several times to get the animal to walk backwards, as it is a clumsy maneuver for grasshoppers due to their long hind legs. Can you tell what sort of stepping pattern is used in backward walking?

Time

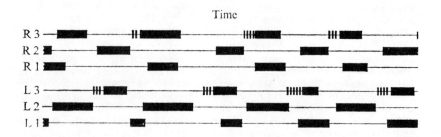

Fig. 8-2. Diagrammatic illustration of the stepping pattern of the 6 legs of a grasshopper walking at about 4 cm/sec. The dark bars indicate the time each leg spends in the air (return strokes). Dashes preceding the return strokes of the hind legs indicate that the animal was dragging its hind legs along the ground before lifting them in the air. (From Burns, 1973)

Lateral Walking. Hold or mount your wooden pole upright and place an animal on it, head up and about half way up the pole. Shine the light from the top, initially. Give the grasshopper a few minutes to adjust to this situation, then move your hand close to the animal. In nature, grasshoppers frequently "hide" from a perceived threat by placing the stem of a plant on which they are resting between themselves and the stimulus. They do this by walking laterally around the circumference of the stem. They will do this to avoid your approaching hand. The stepping pattern they use to walk laterally has never been formally studied. Can you detect a pattern to the stepping?

Metachronal Waves. Although grasshoppers use the tripod or quadrupedal stepping patterns when they are walking at normal rates of speed, this relationship between legs is less obvious at slower walking speeds. Hold the spotlight at different distances from the animal when it is walking straight. Use this technique to vary its walking speed. At slower speeds, the stepping pattern is more spread out in time. Can you discern any patterns in how the legs are used?

You should notice that at slow speeds, 2 points are always true:
1) Legs within a segment alternate in stepping (as is the case with the tripod and quadruped stepping patterns), and
2) the stepping progresses from back to front, with no leg on any 1 side stepping until the 1 behind it has taken up a supporting position on the ground. The anterior progression of this stepping is termed a metachronal wave -- a wave which changes successively in time. This metachronal wave can also be quite readily seen by watching the legs of millipedes and centipedes as they walk.

More detailed information about walking by grasshoppers can be obtained from Burns (1973).

Further Studies of Rotation (Optional). The paths of walking insects in a natural setting are rarely straight lines. Most often the complex nature of the ground requires zigzagging to get from 1 point to another. Body movements under these circumstances have 2 basic components in the plane of locomotion: linear movements and rotational movements. You may have already seen some of the latter in your attempts to get your animal to walk backwards. These body movements form a continuum extending from straight walking at 1 end to pivoting in place at the other. Along this continuum grasshoppers can intermix linear and rotational body movements by alterations in the patterned movements of their legs. You have already seen the leg movements of linear locomotion. Now look at curvilinear locomotion with the aid of an optomotor device.

The optomotor responses of animals are a a group of behaviors which are induced experimentally by a systematic and continuous movement of an animal's visual environment. Under these circumstances animals respond with body movements that minimize the movement of the environment relative to their bodies. (See Exercise 5). For example, a grasshopper which is placed in the stationary center of a drum containing moving black and white vertical stripes will turn in the direction which the stripes move. You can use this behavior to elicit rotational locomotion.

Place an animal in the center of an optomotor drum and begin rotating the stripe pattern slowly. As the animal begins turning, attempt to determine which leg it begins turning with. In most cases it will begin with the foreleg of the side to which it turns. Are all legs used to turn? Do they all walk in the same direction?

At the time of this writing, very little is known about how legs are used for rotation. It is known, however, that grasshoppers can turn in place by walking forward with legs on the outside of a turn and backward with legs on th inside of the turn. This is the same basic way in which a bulldozer turns. Furthermore, these animals can use either the front 4 or all 6 legs to pivot, as they can do in straight walking. For a detailed study of turning in place or "pure" rotational locomotion by an orthopteran insect see Franklin et al. (1981). Can you speculate as to how such turning behavior might have evolved?

C. Jumping

Place a healthy adult grasshopper on a level surface or on a wooden pole and move your hand toward it suddenly. The animal will leap, usually in an undirected way -- not so much toward a distant object as away from the perceived threat. You can observe the actions of the hind legs as they prepare for a jump by moving your hand somewhat more slowly toward an animal, then stopping a short distance away. The grasshopper will prepare to jump by holding the hind legs in a particular way. Describe.

A very interesting mechanism for precontraction of the large tibial extensor muscles in the hind femora has been discovered in grasshoppers (see Heitler, 1974 for details). The femur/tibia joint contains a locking device which allows the relatively small tibial

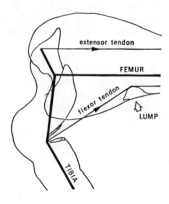

Fig. 8-3. A schematic diagram of the locking mechanism in the hind legs of grasshoppers. The tendon of the femoral flexor is bifurcated for some distance from its attachment to the tibia. When the tibia is fully contracted each side of that tendon nestles up against the lump on the ventral interior surface of the femoral cuticle. Further contractions of the flexor muscle tighten the tendon around the lump and by holding the tibia in place prevent the much stronger extensor muscle from extending the tibia. (Slightly modified from Heitler, 1974)

flexor muscle on the ventral side of the femur to hold the hind tibia fully flexed while the much more powerful extensor muscle is contracted (Fig. 8-3). In this way the extensor muscle is readied for the jump before the necessity of its execution. The jump occurs when the locking mechanism of the flexor muscle is released, allowing the extensor to move the tibia with an explosive burst of power.

This locking mechanism can be inactivated by preventing the hind tibiae from flexing fully. To do so, place a piece of wax on the ventral surface of the femur just anterior to its articulation with the tibia. Now try to make the grasshopper jump.

Peering. Prior to jumping toward an object, grasshoppers perform an interesting behavior called "peering". These lateral oscillations of the front part of the body allow them to determine the distance between themselves and the objects to which they wish to jump (Wallace, 1959). This behavior is not evident when grasshoppers are jumping to escape danger.

A ring stand, wooden pole, or other object resembling the stem of a plant will serve as a stimulus for this behavior. Darken the room and illuminate the top half of the "stem". Place the grasshopper on a small box, book, or other platform which elevates it several inches from the ground, and face it towards the illuminated object. Watch carefully as soon as you release the animal; it will gauge the distance to the "stem" very quickly. Can you figure out how this behavior aids the animal in determining distance?

Coordination of Legs During the Jump.   Straight   walking   is
characterized   by   alternation   in   the   stepping   of   legs   within   a
segment.   Clearly,   jumping   by   kicking   with   hind   legs   in   alternation
would   only   provide   half   the   thrust   of   kicking   with   2   in   synchrony.
Since   jumping   evolved   after   walking,   the   walking   pattern   was   modified
so   that   legs   moved   together   rather   than   in   alternation.   Thus,   the
hind   legs   kick   in   unison   and   the   front   and   middle   legs   splay   apart,
moving   synchronously   as   the   animal   sails   through   the   air.   (The
posture   of   these   legs   is   thought   to   prepare   the   animal   for   landing.)
Observe this for yourself.

D.   Swimming

Grasshoppers   often   must   jump   in   an   undirected   way   to   escape
potential   predators.   In   areas   where   there   are   bodies   of   water   they
frequently   land   in   the   water.   This   is   an   extremely   dangerous
situation   for   them,   as   they   can   quickly   become   fish   food.   Thus   it   is
not   surprising   to   find   that   grasshoppers   have   evolved   swimming
movements of their legs from their jumping movements.

Initial Orientation Movements.   When   a   grasshopper   jumps   into
the   water,   the   first   problem   it   encounters   is   finding   the   shore   in
order   to   swim   toward   it.   In   performing   this   orientation   it   executes
a   complex   series   of   leg   movements   designed   to   turn   its   body   in   the
water.   You   can   simulate   these   conditions   by   placing   a   vertically
oriented   pattern   of   black   and   white   stripes   on   1   side   of   a   large
white   porcelain   bowl   filled   with   lukewarm   water.   Place   a   grasshopper
in   the   water   facing   at   some   angle   to   the   stripes   (most   of   the   time).
How does it manage to turn?

Two   basic   kinds   of   turns   can   be   distinguished   by   the   way   the
hind   legs   operate.   Moderate   turns   are   characterized   by   the   animal
kicking   with   the   hind   leg   away   from   the   turn   direction   (Fig.   8-4);
this   is   like   turning   a   rowboat   by   rowing   with   1   oar.   Sharp   turns
involve   the   use   of   both   hind   legs   (Fig.   8-5).   The   tibia   of   the   leg
on   the   side   of   the   turn   is   held   perpendicular   to   the   long   body   axis
to   act   as   a   drag   or   pivot,   and   the   leg   away   from   the   turn   kicks   to
provide   propulsion.   Occasionally   your   animal   may   even   row   backward
with the tibia on the inside of the turn.

Straight Swimming.   Upon   successful   orientation   to   the   stripe
pattern,   your   grasshopper   will   begin   to   swim   straight   toward   the   side
of   the   bowl.   To   do   so,   it   performs   synchronous   kicking   movements
with   its   hind   legs   (Fig.   8-6).   Is   this   movement   the   same
as   jumping?   Jumping   requires   the   use   of   the   locking   mechanism
previously   discussed.   Therefore,   1   way   to   find   out   whether   they   are
using   the   same   motor   pattern   is   to   block   the   precontraction   mechanism
as before.   Is the grasshopper still able to swim?

Notice   the   movements   of   the   fore   and   middle   legs   during   normal
straight   swimming.   They   move   together   (in   unison),   and   the   spreading
or   splaying   movements   are   quite   similar   to   those   performed   for   jumps.
The   hind   femora   are   not   held   in   quite   the   same   position   relative   to
the   body   as   they   are   for   jumps.   For   swimming   they   are   located
parallel   to   the   abdomen;   for   jumps,   they   are   elevated   at   an   angle

Moderate   Turning

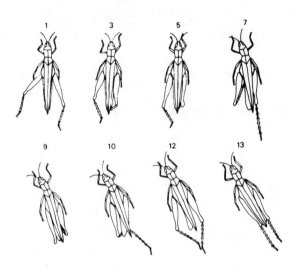

8-4. The sequence of leg movements made to execute a moderate turn. <u>Melanoplus</u> <u>differentialis</u> filmed at 50 fr./sec. Numbers above the figures represent the frame number from the onset of the turning behavior. (From Franklin et al., 1977)

Sharp Turning

Fig. 8-5. Leg movement sequence for sharp turns. Numbers as in Fig 8-4. (From Franklin et al., 1977)

Straight Swimming

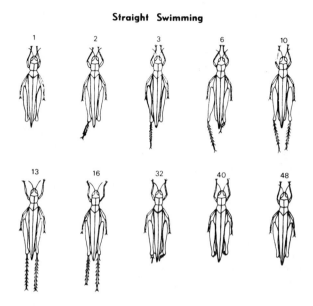

Fig. 8-6. Leg movements for straight swimming by <u>Melanoplus</u> <u>differentialis</u>. Numbers as in Fig. 8-4. (From Franklin, et al.,1977)

relative to the body. Can you think of a mechanical reason why a grasshopper assumes these different positions with its hind legs?

Mechanics of Swimming. Only the hind legs contribute to a grasshopper's propulsion in water. Verify this by waxing the animal's fore and middle legs to its thorax or by amputating them. In fact, a grasshopper can swim with a single (hind) leg. How is such propulsion accomplished? The tibiae of the hind legs are the active elements of propulsion; animals without them cannot swim at all. However, the tibiae are basically round poles being moved back and forth in the water. They should provide equal thrust on being extended from the body (a power stroke) or on being flexed in a return stroke. Watch the actions of the hind legs during swimming and see if you can guess what the propulsive mechanism is.

Observing your grasshopper carefully, you might have noticed that extension of the tibiae is more rapid than their return or flexion. In fact, it is several times faster (though the extension is much slower than it is during jumps on land). The resistance of the water to the moving tibiae, and thus the resistance against which the tibiae can push, is a function of the square of their velocity through the water. Therefore, if the tibiae travel more rapidly during their extension than flexion, there is more push to generate forward movement than the subsequent detracting pull due to the slower return. In 1 species of grasshopper, the tibiae extended 4 times faster than they were flexed; this resulted in 16 times more force during the power stroke than the return stroke (Franklin et al., 1977).

Differences in the extension and flexion speed of the hind tibiae appear to be behavioral adaptations of jumping to take advantage of the hydrodynamic properties of water (Franklin et al., 1977). Physiological confirmation of the adaptation of the jumping motor pattern for swimming has recently been provided by Pflüger and Burrows (1978).

SUMMARY QUESTION

1. Grasshoppers pivot on the spot by walking backward with legs on the inside of the turn and forward with legs on the outside. Is it physically possible for these animals to pivot by coordinating their legs with the tripod pattern that they normally use when legs on both sides of their body are walking forward -- as for straight walking?

SELECTED REFERENCES

Bennet-Clark, H.C. 1975. The energetics of the jump of the locust Schistocerca gregaria. J. Exp. Biol. 63: 53-83.
Brown, R.H.J. 1967. Mechanisms of locust jumping. Nature, London 214: 939.
Burns, M.D. 1973. The control of walking in orthoptera. I. Leg movements in normal walking. J. Exp. Biol. 58:45-58.
Chapman, R. F. 1971. The Insects: Structure and Function. 2nd. ed. American Elsevier Publishing Co., New York.
Franklin, R., R. Jander and K. Ele. 1977. The coordination, mechanics and evolution of swimming by a grasshopper, Melanoplus differentialis. J. Kansas Entomol. Soc. 50: 189-199.
Franklin, R., W. J. Bell, and R. Jander. 1981. Rotational locomotion by the cockroach Blattella germanica. J. Insect Physiol. 27: 249-255.
Gray, Sir James. 1968. Animal Locomotion. Weidenfeld and Nicholson, London.
Heitler, W.J. 1974. The locust jump. J. Comp. Physiol. 89: 93-104.
Hughes, G.M. and P.J. Mill. 1974. Locomotion: Terrestrial. In M. Rockstein (ed.), The Physiology of Insecta, vol. 3, 2nd ed. Academic Press, New York, pp. 335-379.
Manton, S.M. 1977. The Arthropoda; Habits, Functional Morphology and Evolution. Claredon Press, Oxford.
Nachtigall, W. 1974. Locomotion: Mechanics and hydrodynamics of swimming in aquatic insects. In M. Rockstein (ed.), The Physiology of Insecta, vol. 3, 2nd ed. Academic Press, New York, pp. 381-432.
Pflüger, H.J. and M. Burrows, 1978. Locusts use the same basic motor patterns in swimming as in jumping and kicking. J. Exp. Biol. 75: 81-93.
Wallace, G.K. 1959. Visual scanning in the desert locust Schistocerca gregaria. J. Exp. Biol. 36: 512-524.

9. Optimal Foraging Strategy: I. Flower Choice by Bumble Bees

Bernd Heinrich
University of Vermont

Bumble bees are known to visit flowers exhibiting a wide range of colors, morphologies, and food rewards. Since most colonies in temperate regions last from 2 to 5 months, while many species of plants bloom for only 1 to 2 weeks, the bees of a single species must utilize, both locally and temporally, many different kinds of flowers for their hive economy.

While foraging, bumble bee workers are remarkably "single-minded", being undistracted by predator avoidance behaviors, by aggressive interactions, or by sexual behavior. Furthermore, their foraging is often energy-limited. Therefore, they might be expected to forage "optimally" -- that is, to behave in such a way that they maximize their rate of energy gain per unit of foraging time. This implies that they will move through a productive flower patch in a way that increases their rate of encounters with preferred foods, and that they will forage preferentially from the flowers where they can make the most rapid profits.

Many factors are probably involved in flower choice. The quantity and type of floral reward is one. Some of the kinds of flowers provide only pollen. Others provide only nectar. Many provide both pollen and nectar, or neither. Furthermore, the amounts of these rewards per flower can vary by an order of magnitude.

Flower abundance is certainly another factor. Most foragers visit many flowers per minute, and the net food rewards available per unit time depend not only on the amount and quality of reward per flower, but also on the absolute number of flowers that can be visited per unit time.

Flower morphology is also important, for it affects access to the nectar and/or pollen. Different kinds of flowers require different "handling skills" to harvest their rewards. For example, many typical "bumble bee flowers" have their nectar at the bottom of a deep corolla tube. In order to extract it, the bees have to enter the flower at a specific entrance that may, in some cases, have to be pried open. On the other hand, many flowers such as members of the Compositae provide both pollen and nectar in numerous tiny florets arranged on an open landing platform; these rewards, though usually minute, are available even to foragers with minimal foraging skills.

In this exercise we will examine the foraging behavior of bumble bees as it relates to flower choice; in the exercise that follows (Exercise 10), we will examine foraging as it relates to changes in reward availability.

METHODS

Subject

Bumble bees, species of <u>Bombus</u>, can usually be recognized by their robust shape, size (most being 19 mm in length or more), and black and yellow coloration; a few are marked with orange. There are approximately 50 bumble bee species from Alaska to Mexico. Some species are widespread. In any 1 local area, there may be up to 6 common species of bumble bees. The color and range guide in the appendix of Heinrich, 1979b (see Selected References) provides bee identification of information.

Because the colony cycle is annual (Fig. 9-1), in spring and early summer all bumble bees encountered are queens. In late summer, bumble bees are usually extremely common but almost all the bees one sees are workers. In the fall, drones will also be foraging. Additional information on bumble bee behavior and ecology may be found by consulting the Selected References.

NOTE: Students hypersensitive to bee stings should not participate in this exercise.

Materials Needed

Insect nets; 1 µl and/or 5 µl capillary tubes (such as Drummond Microcap); a refractometer; bee identification guide; stopwatches; mm ruler; 30-100 wire rods or thin poles with flagging; tape measure; ethyl acetate killing bottle. Optional: table sugar (sucrose); eye droppers.

Procedure

Some of the observations can be made by individuals working by themselves. Others will require a team approach. The data will be pooled among the participants at the end of the experiments.

As a study area, choose a meadow, bog, prairie, or some other relatively open area where there are a variety of plants blooming at the same time, and where bumble bees are present.

A. Individual Behavior at Flowers

Observe bees closely at as many different kinds of flowers as possible. What are the details of how each flower is "handled"?

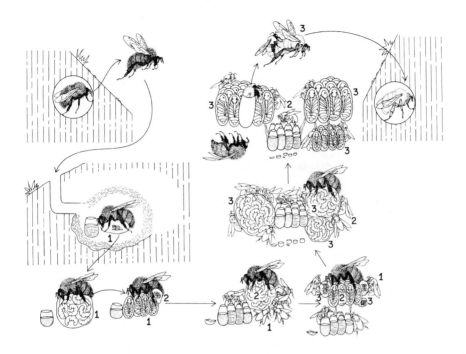

Fig. 9-1. Diagram of a bumblebee colony cycle, from the queen emerging from hibernation (left) to new queens (lightly stippled) emerging from cocoons of the third brood (eggs at lower right), mating and hibernating (right). (From Heinrich 1979b)

How many times does the bee probe for nectar in each flower? Where does the bee probe? How is pollen collected and disposed of? Are there individual differences in handling behavior at the same kinds of flowers

B. Resource Partitioning

Working in pairs, follow a single bee for 5 min, collecting each flower it visits. Collect the individual and kill it in an ethyl acetate killing jar, which usually will cause its mouthparts to extend. Measure tongue length to the nearest mm. Identify the bee and record body size.

Measure lengths of the corolla tubes of the flowers which it visits. Is there any correlation you can discern? Draw the details of the morphology of each flower type. Note the location of the nectar and pollen.

Open the flowers carefully. Withdraw the nectar into a dry capillary tube and measure the volume, using the mm rule. Nectar volumes per flower are often minute. In order to extract the nectar you may have to cut off the top of the corolla tube and squeeze the nectar droplet up out of the flower, using your thumb and forefinger.

After you have collected 1 - 2 µl of nectar, deposit the droplet into the refractometer to read sugar concentration (wt of sugar/total wt of sample).

Repeat entire procedure for several individual bumble bees, of different species if possible, as time permits. Combine pooled class results for the different species. Is there evidence of resource partitioning among the species present?

C. Resource Specialization

Do the bees display any preferences for particular kinds of flowers out of the choice available? To approach this question experimentally, 2 or 3 persons should work as a team. One or 2 observers should follow a single bee, carrying a bundle of 30-100 consecutively numbered flags on wire rods or poles and a stopwatch. Another should act as recorder. In repeat trials, switch roles.

Having chosen a bee to follow, call out the identity of each flower as the bee under observation lands upon it. The recorder should enter the name on a running list, with duration of the foraging episode on that flower. After the bee leaves the flower, position a stake with a numbered flag near that flower. Follow the bee for as long as possible. (Particularly among very dense patches of flowers, there may be intervals when it is impossible to keep accurate track of foraging duration on all individual flowers. However, in these cases you should still be able to work backwards and calculate number of flowers visited per time interval.)

At the end of observations on this bee, remove the flags in sequence. Record distances between flowers visited. Tabulate the kinds and number of flowers on and to about 1 m on each side of the flight path. Compare the kinds of flowers available to the bee with those it visited. Did the bee pass over all kinds of flowers equally? Did it display a preference out of those available? How are the data biased? How might one reduce some of this bias?

Pool your data from parts A and B with the class, tabulating the different flowers and manipulative behaviors observed in each bumble bee species.

SUGGESTIONS FOR FURTHER STUDY

Time and circumstances permitting, you may also wish to do other experiments to supplement your observations.

1. Foraging Profits

The food rewards -- nectar and/or pollen -- of flowers vary a great deal. Examine the potential foraging profits that can be collected from the various available flowers in an area, and relate this to the number of bees you can count utilizing these various flower species. Nectar can be collected as outlined earlier in this experiment. Refractometer readings can be used to determine calories of food energy available. One µl equals close to 1 mg; 1 mg sugar equals approximately 3.7 calories. Compare the energy rewards with the values of energy expenditure published in the references cited at the end of this exercise.

It is generally thought that bees collect pollen secondarily to collecting nectar. However, some flowers are visited exclusively for their pollen. Pollen rewards are very difficult to measure quantitatively in the field. Little work has been done on this subject. In the laboratory you might wish to shake the pollen out onto pre-weighed paper and weigh it.

2. Flower Enrichment

Make a 20-30% sucrose solution. With an eye-dropper, place droplets of this solution onto or into flowers where bees are foraging. What changes in behavior do you observe?

3. Flower Alteration

Do the bees respond to the flower signals (color, scent, geometric patterns), to the food rewards, or both? In a population of flowers where bees are foraging, remove the petals of half of the flowers, or scent half of them with a different scent (anise, perfume). Record the visits to both kinds.

4. Artificial Flower "Field"

Create a temporary meadow, using a few hundred containers and a source of different flowers. Set out jars or cans with water in a large grid and insert into each an identical bouquet of several different kinds of flowers. This procedure has the advantage that you can introduce "novel" flowers from a different area that the bees may not have contacted previously into an area that already has an established set of flowers. How do the bees react to these new flowers?

SUMMARY QUESTIONS

1. In what ways do bumble bees satisfy the conditions for an optimal forager? Do they appear to forage preferentially from the flowers where they can make the most rapid profits? What is your evidence?

2. Characterize "preferred food" for your most commonly encountered bumble bee species. How does this species increase its rate of encounters with preferred food? (Caution: What circular reasoning is a danger in answering this?)
3. Characterize the flower morphology types most commonly used by your bumble bees. What pattern emerges? Were you able to correlate this to bee morphology? How?

4. By what mechanisms does intraspecific competition for flower resources appear to be minimized among bumble bees?

5. How might flower choice differ for long- and short-term

optimality?

6. What are the primary cues used in determining flower choice?

SELECTED REFERENCES

Brian. A.D. 1957. Differences in the flowers visited by four species
of bumblebees and their causes. J. Anim. Ecol. 26: 71-98.
Emlen, J. M. 1968. Optimal choice in animals. Amer. Nat. 102:
385-390.
Grant, V. 1950. The flower constancy of bees. Bot. Rev. 16:
379-398.
Heinrich, B. 1972. Temperature regulation in the bumblebee, Bombus
vagans: a field study. Science 175: 183-187.
Heinrich, B. 1973. The energetics of the bumblebee. Sci. Amer. 228:
96-102..
Heinrich, B. 1976. Foraging specializations of individual bumblebees
Ecol. Monogr. 46: 105-128.
Heinrich, B. 1979a. "Majoring" and "minoring" in bumblebees: an ex-
perimental analysis. Ecology 60: 245-255.
Heinrich, B. 1979b. Bumblebee Economics. pp. 215-220, 246-249.
Harvard University Press, Cambridge.
Hobbs, G.A. 1962. Further studies on the food-gathering behavior of
bumblebees. Can. Entomol. 94: 538-541.
Manning, A. 1956. Some aspects of the foraging behavior of bumble-
bees. Behaviour 9: 164-201.
Pyke, G.H., H.R. Pulliam and E.L. Charnov. 1977. Optimal foraging
theory: A selected review of theory and tests. Quart. Rev.
Biol. 52: 137-154.
Witham, T.G. 1977. Coevolution of foraging in Bombus and nectar-
dispensing in Chilopsis: a last dreg theory. Science 197:
593-596.

10. Optimal Foraging Strategy: II. Foraging Movements of Bumble Bees

Bernd Heinrich
University of Vermont

The rate at which a bumble bee colony can grow is limited by the food supplies that the workers can bring into the colony (Fig. 10-1). The workers' sole activity when in the field is to forage; they do not alter their movements for functions other than those required to harvest food. One might expect, therefore, that they forage "optimally" -- that their movements are dictated primarily to enhance the rate of food intake.

The distribution of food in the environment is varied, and there are many different potential options in foraging movements. Flowers tend to be patchily distributed, and some patches contain much larger food rewards per flower than others. A bee can never know which is the "best" patch unless it has sampled many patches. But the more the bee samples different patches, the more time and energy it may waste in potentially fruitless search -- time and energy that might have been more profitably spent foraging.

After a bee has restricted its foraging, at least temporarily, to a specific patch, it could visit closest flowers, or it could bypass some flowers. By always visiting closest flowers the bee might soon be re-visiting flowers it has just emptied. However, by bypassing many flowers the bee might soon find itself outside the patch. Individual inflorescences with many florets can also be considered as mini-patches, and the same alternatives would apply. How might the different options vary with differences in the amount of food per flower or floret? Why?

The purpose of this exercise is to determine whether or not bees make adjustments in their foraging movements in response to changes of reward availability. You will try to determine the "rules" by which the bees regulate their foraging movements on small and large patches of flowers with varying amounts of food per flower and/or with varying flower density.

METHODS

Subject

You can use any of a large variety of insects as experimental

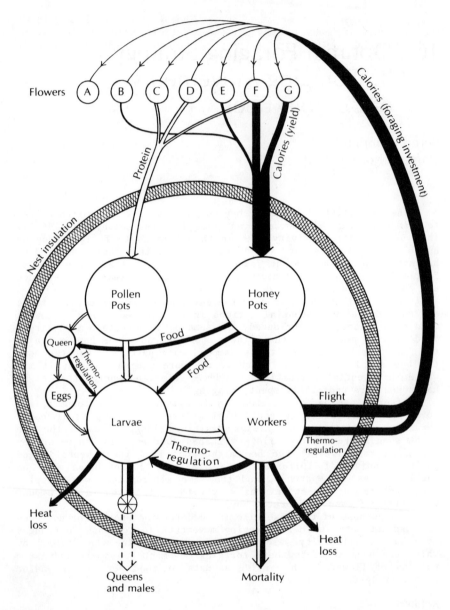

Fig. 10-1. Flow diagram showing movement of matter (protein) and energy (calories) from the environment and through a bumble bee colony. (From Heinrich 1979b)

subjects. However, either honey bees or bumble bees are likely to be the most suitable because they are the least disturbed and distracted while foraging. They are also very common in almost any area suitable for the observations you will be making.

Clover (Trifolium) is a common lawn "weed" whose flowers are actually dense heads, or inflorescences, each comprised of many small florets. Bees -- both honey bees (especially near an urban area) and bumble bees -- are important pollinators of clover. Other locally abundant flowers attractive to bees could be substituted, with minor modifications.

NOTE: Students hypersensitive to bee stings should not participate in this exercise.

Materials Needed

300 m white string; 160 stakes; scissors or jackknife for cutting string; small mesh size screening, at least 3 m square (dacron or nylon marquisette or window screen); graph paper; rulers; protractor; 1 µl and/or 5 µl capillary tubes (such as Drummond Microcap); refractometer.

Procedure

A. Foraging and Nectar Concentration

Three days prior to class, your instructor will have staked out 2 study plots, 3 m². Working in teams, measure the nectar concentration of a sample of the flowers from both plots, using the procedures outlined in the previous exercise (Exercise 9). Remove the screening from the experimental plot. Estimate and record the average flower density in the 2 plots.

Working individually, observe bees that enter the 2 plots. (Several persons can make simultaneous observations, each following different individual bees.) Record the distance bees move between successively-visited flowers, the number of florets visited per inflorescence, and the turning angles from 1 inflorescence visited to the next. (It will probably not be possible to keep track of the actual flight paths.)

To aid you in recording these data, you can use the grid lines laid out in the flower patch. Reproduce the grid lines on a sheet of paper. Find the corresponding grid square on the experimental plot where you first locate your bee. Draw a dot on your paper with grid lines to mark the bee's approximate location on the plot. When the bee moves to the next inflorescence, mark that location by another dot and connect the 2 points. Continue this procedure for as long as the bee remains on your plot.

Repeat your observations and mapping for additional bees as time permits, alternating between the experimental and control patches. From your flight maps, measure the distances of the individual moves relative to the 15 cm grid lines. Turning angles at each inflorescence can be measured with a protractor, relative to the straight lines drawn between the 2 last-visited inflorescences. Plot separate graphs (histograms) of the frequencies of the different inter-inflorescence distances, angles, and number of florets visited. Compare the 3 histograms of the data from the low-nectar patch with the data from the high-nectar patch.

B. Foraging and Reward Density

As an alternate or supplement to the above experiment, you may wish to compare the foraging movements between 2 patches with different inflorescence densities, rather than different rewards per inflorescence. You can manipulate flower-head densities by plucking flowers from a patch until the desired density is achieved. The 2 patches should differ by a factor of at least 5 times. You may want to have the sparsely-flowered patch of a larger size.

SUGGESTIONS FOR FURTHER STUDY

To determine whether individual bees adjust their foraging strategies in response to changes of reward availability, follow specific individual bees over many consecutive foraging trips. (Some bees may forage in the same area over hundreds of consecutive foraging trips.) This requires marking for individual recognition. There are at least 3 potential methods:

1) Anesthetize bees in ether (but only _very_ lightly) and glue the commercially-available Opalithplattchen onto the thorax. An advantage of this method is that individuals are relatively permanently marked, and there is little or no ambiguity with the discrete numbers and colors. The labels are easy to read. A major disadvantage is that the bees are disturbed by the marking. Most will not be seen again after they are released.

2) Hold bees fast and paint the thorax with typing-eraser fluid (such as Liquid Paper). It dries quickly. The paint patch can be marked with a number using a rapidograph pen. The advantage of this method is that it is quick and inexpensive. The paints dry quickly and the bees do not need to be anesthetized to keep them from removing the marker. The disadvantage is the same as for the previous method.

3) While they are on flowers, dab bees with paint, such as Testor's quick-drying butyrate dope. Bees can be marked without interrupting their foraging behavior, or they are only minimally disrupted. However, some of the paint often wears off after a few days, and it is difficult to have many individually distinctive marks if one wants to keep track of large numbers of individuals.

SUMMARY QUESTIONS

1. Do the bees visit the closest flowers in succession in both the low and high reward situations?

2. Does the bee's direction of movement from 1 inflorescence to another vary with the food rewards encountered on previous inflorescence(s)?

3. How do the answers.to the above 2 questions relate to movements on the inflorescence itself, and vice versa? The dense head of florets which comprises a single clover flower might be considered a "mini-patch". What evidence is there that bees treat it as such?

4. What is the difference in the percentage of flowers approached and then bypassed without landing in the low and high reward situations? Does this vary with flower-head density?

5. Explain your above results in terms of foraging behaviors you might predict if the rate of food intake were to be maximized.

SELECTED REFERENCES

Emlen, J. 1966. The role of time and energy in food preferences. Amer. Nat. 100: 610-617.

Heinrich, B. 1979a. Resource heterogeneity and patterns of movement in foraging bumblebees. Oecologia 140: 235-245.

Heinrich, B. 1979b. Bumblebee Economics. Harvard University Press, Cambridge, MA.

MacArthur, R.H. 1966. On optimal use of a patchy environment. Amer. Nat. 100: 603-609.

Pyke, G.H. 1978. Optimal foraging movement patterns of bumblebees between inflorescences. Theor. Pop. Biol. 13: 72-98.

Pyke, G.H., H.R. Pulliam and E.L. Charnov. 1977. Optimal foraging theory: A selective review of theory and tests. Quart. Rev. Biol. 52: 137-154.

Schoener, T.W. 1971. Theory of feeding strategies. Annu. Rev. Ecol. Syst. 2: 269-404.

Waddington, K.G. and B. Heinrich. 1979. The foraging movements of bumble bees on vertical "inflorescences": an experimental analysis. J. Comp. Physiol. 134: 113-117.

11. Foraging Behavior of Subterranean Termites

William L. Nutting
University of Arizona

Termites live in large long-lived colonies, fully comparable to ant societies in their organization and complexity. They are primarily a tropical group, but occur in all states except Alaska, although we rarely see them or evidence of their activity. Most of the common and destructive species are subterranean -- soft, delicate, unpigmented insects, generally blind, and easily desiccated. The colony lives in a system of underground chambers and galleries, a relatively stable and secure closed environment opened only to release winged reproductives during the annual flight season.

Cellulose is the most abundant of the continuously cycled organic compounds in the biosphere. Woody litter (logs, branches, twigs, and other plant debris) continually falls to the ground, accumulating at annual rates which are impressive even in semi-arid ecosystems. The major consumers of this woody litter are termites, fungi, and other microorganisms.

Termites are detritivores and herbivores whose ability to utilize cellulosic materials efficiently has almost certainly contributed to their widespread success. Yet we are only beginning to realize how large a role they actually play in processing dead plant material and in nutrient cycling. Studies of termite foraging behavior have barely progressed beyond the observational stage. However, there are techniques which show promise for an experimental approach to the subject (LaFage and Nutting, 1977).

Considering our ignorance of these legendary insects, we can begin with a very general question: How do these blind and "insignificant" insects find their food? More specifically, what are the foraging strategies of subterranean termites? What cues do they use to locate suitable food items on the soil surface above them? Since they appear to live in a world of darkness, what regulates their activities? Do they follow temperature or moisture gradients, home in on chemical cues? Is their foraging circumscribed within "territories", limited by temperature, modulated by food preferences

I acknowledge the full participation of M.I. Haverty and J.P. LaFage in the development of this method of studying subterranean termites.

and/or colony needs?

The few existing studies on subterranean termites have been based on baiting with attractive materials, or on soil-core sampling. In this exercise we will use a modification of the bait-sampling method which provides a picture of the size and spatial distribution of their surface foraging populations. It can also provide us with some clues to the factors regulating foraging activity and insight into their food-finding strategies (LaFage, Nutting and Haverty, 1973). Plain white, unscented toilet paper rolls have proven to be easily handled, uniform bait units. They are set out in a regular pattern on the ground as a food source which brings the termites to the surface where their foraging activity can be monitored on a continuing basis.

This exercise is a long-term study. It requires a few weeks at the very least, and could profitably be run for a year or the length of a foraging season. After a suitable study area is selected and set up, the rolls or other baits are examined weekly; more frequent disturbance is likely to inhibit foraging activity.

METHODS

Subject

Out of some 2200 species of Isoptera, about 40 species occur in the United States (Snyder, 1954; Weesner, 1965). Six look-alike species of Reticulitermes occupy overlapping ranges that cover most of the United States. Any of these could probably be used. Reticulitermes tibialis has been studied on toilet paper grids in Utah and New Mexico; R. flavipes has been used in the Southeast. One other member of the Rhinotermitidae, Heterotermes aureus, is common in arid to semi-arid areas of southern Arizona and southeastern California, and comes readily to toilet paper.

About a dozen species of the Termitidae occur over varying ranges in the Southwest. Most of the 8 species of Amitermes appear to have limited ranges, although this may be simply because we know very little about them. A. wheeleri and 2 species of Gnathamitermes are widespread and fairly common. They are also attracted to toilet paper.

Of the 17 U.S. dry- or damp-wood-inhabiting species, only 1 is likely to attack paper or wood on or in the soil. This is Paraneotermes simplicicornis, a kalotermitid that attacks stumps, roots, and buried wood across the Southwest from southeastern California to southwestern Texas. We have seen it in toilet rolls in southern Arizona grassland.

Materials Needed

Toilet paper, case of 100 roles, plain white, unscented; tape, 2 inches wide, good quality, of the type used for sealing heating-coooling ducts (should be waterproof); metal stakes, "rebar", for grid corners; garden rake, other tools for clearing litter from plot; measuring tape, steel, at least 10 m. <u>Desirable</u>: Thermometer, portable, with thermocouple or thermister probe; thermometer, max/min; rain gauge, wedge type; hygrothermograph, weekly recording. <u>Optional</u>: Iron wire, #4, 25 m; aluminum foil, sufficient to cover top and circumference of the rolls used, <u>or</u> plastic bags, size and number appropriate to enclose number of toilet rolls used.

Procedure

Your instructor will have chosen a suitable study location. As a class, your roles include preparing the site, preparing and setting out bait units, and examining the rolls on a convenient predetermined schedule. At some point, termite activity will be followed through a day and a night.

A. Preparing the Bait Units

The toilet rolls (or alternative bait units) may be readied well in advance of the field work. They should at least be wrapped circumferentially with tape or, in areas of heavy rainfall, covered (except for the bottom) with aluminum foil or plastic bags. If necessary, the rolls can be held down with 25 cm lengths of #4 iron wire bent over at the top like a European 1. These should be cut and bent ahead of time.

B. Preparing the Study Site

After the site has been selected and shortly before setting out the bait units, the plot should be cleared of woody litter. This is likely to be the most time-consuming part of the project. However, it is not necessary to remove living plants or dead leaves and fine litter. Heavily wooded areas and large logs should be avoided, although nearby clear areas might be very suitable.

The bait units should be set out as soon as possible after clearing the site, preferably on the same or the next day. Since the termites have been deprived of their usual food items, they are likely to begin exploratory activity. With the materials assembled, and the rolls wrapped, 2 people with a steel tape can lay out and set up the suggested grid in a couple of hours.

Lay out a rectangle; 10 m x 10 m is a good size and, in most areas, probably large enough. The corners should be marked with flagged metal stakes, but meter intervals of the grid can be marked with sticks or pebbles. The wrapped toilet rolls, 100 in all, can now be set out on the grid. They should be worked into the surface of the soil so as to insure good contact.

C. Monitoring Termite Foraging Activity

Some convenient schedule for examining the rolls must be worked out among class members. Ideally the baits should be disturbed as little as possible; once a week is reasonable. With a little practice, 100 rolls can be checked in about 30 min.

Actual checking for termite foraging activity involves picking up each roll in succession, quickly examining the underside, and on the data sheet (Fig. 11-1) noting the species and approximate number of individuals present or, if none, any evidence of past activity. Rapid gross estimates of numbers of small, moving insects can be made by using a series of categories: 1-5; 6-50; 51-150; 151-250; >250. The accuracy of these estimates could be verified (and corrected if necessary) against some actual counts at the end of the study.

A few physical factors should be measured in the vicinity of the plot to accompany the biological data: air temperature (daily max/min), precipitation (weekly totals), and soil moisture (gravimetrically at 15 cm) if possible. An example of such data is shown in Fig. 11-2. Additionally it would be useful to take temperatures (thermistor probe) inside a particular roll, at the roll-soil interface, and 5 or 10 cm below the roll, and to compare them with concurrent readings of the air, soil surface, and at the same soil depth 15-20 cm away from the roll. Although not necessary, a recording hygrothermograph maintained on or near the site would provide a record from which maxima, minima, means, and other useful parameters could be calculated as desired. The following references will provide useful background here: Beard, 1974; Haverty, LaFage and Nutting, 1974; Haverty, Nutting and LaFage, 1975; LaFage, Haverty and Nutting, 1976.

D. Analyzing Your Data

At the completion of this exercise, pool the class data for analysis. Among the questions you will wish to consider are:

1. On your grid are there any particular patterns of attack (foci of intense or long-term foraging) which suggest that certain areas might be foraging territories of individual colonies? (See Haverty, Nutting and LaFage, 1975.)
2. From what you have learned do you believe that these termites are generally distributed throughout your area?
3. If your data are sufficient, can you make reasonable estimates of the following?
      a. Foraging populations/hectare (10,000 square meters).
      b. Paper consumption/hectare/year.
      c. Quantity of soil brought to the surface/hectare/year.
      d. Woody (or other) litter consumption/hectare/year. (See Haverty and Nutting, 1975b).
      e. Percent of litter standing crop consumed.
      f. Percent of annual litter production consumed. (What happens to the rest of the litter?)

DATE _ _ _ _ _ _   SITE -

TEST AREA —   TIME _ _ _ hrs.

SAMPLED BY _____   _____

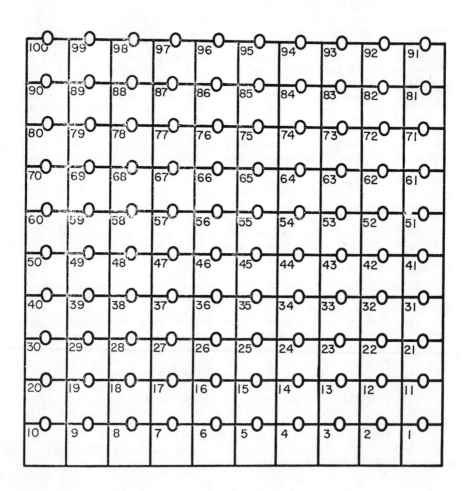

Fig. 11-1. Sample grid for collecting information on termite foraging.

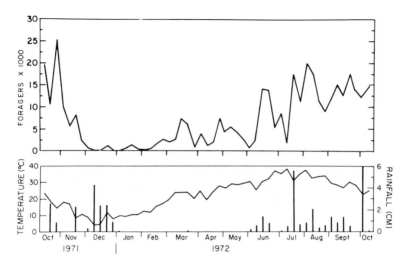

Fig. 11-2. Annual changes in numbers of desert subterranean termite foragers as related to temperature and precipitation. (From Haverty et al., 1974)

SUGGESTIONS FOR FURTHER STUDY

1. Laboratory Investigations of Termite Feeding

    You may wish to study termite behavior further in the laboratory, perhaps using some of the summary questions as your focus.
    Specimens for demonstration or study are best brought back to the laboratory in the wood in which they were found, put into a plastic or heavy paper bag. They can be removed by sawing the wood into short lengths, splitting it along the termite galleries, and extracting individuals tediously by hand. Sometimes enough can be collected by "papering": Lay the wood (split if large) to dry in large enameled or plastic trays, and cover with damp paper towels and a plastic sheet. The termites will usually migrate from the wood to the paper, where they can be brushed off into a culture dish with a small camel's hair brush.
    Subterranean termites are notoriously difficult to keep in the laboratory. The simplest and most effective method is to provide them with good-sized pieces of the wood they were collected in, placed on top of a deep layer of damp, fine vermiculite in an all glass aquarium. It should be covered with a glass plate or aluminum foil, and the vermiculite should be kept damp but never wet. A vigorous group, preferably with reproductives, is the best insurance against takeover by mold. Several hundred to a few thousand individuals in a 5 gallon container is a good number. Vermiculite is preferable to soil since the former contains no organic material

which would support mold. Becker (1969) provides a thorough review of rearing and breeding methods for all kinds of termites.

2. Termite Feeding Preferences

The following types of test have been used for determining the vulnerability of various materials to termite attack in many parts of the world. They probably should not be tried until an area is found with a fairly high, dependable termite population. The experiments should be planned carefully with a view toward obtaining statistically meaningful results, and approved before they are set up in the field.

a. Preferences for cellulosic materials: wood, chipboard, hardboard, celotex, cardboard, paper, cow chips.
b. Preferences for different woods: various native, exotic woods; fungus-infested vs. uninfested wood; dry (covered with plastic) vs. damp wood (watered periodically). (See Haverty and Nutting, 1975a).
c. Attractiveness of non-cellulosic materials: stone, brick, glass, sheet metal, plastic, asbestos. (Termites and many other arthropods are commonly found under stones. For what reasons might termites come up under stones and other inedible objects?)

In comparing the above materials, it is important to present an equivalent surface area of each sample to the soil. In all cases the material should have good overall contact with the soil. For the wood tests, 15 cm squares of 2-3 cm board, weighted with a brick, are satisfactory.

SUMMARY QUESTIONS

1. Not unlike many ants, most of our subterranean termites are believed to have 1 or more "nests" or groups of chambers at varying depths in the ground (sometimes in logs) which they frequent according to season. How do you envision that they explore a territory for new items of food? How do they carry their forage back to the "nest"?

2. What cues do you suspect termites are using to pinpoint new food items on the soil surface?

3. What physical or abiotic factors most influence foraging intensity (number of foragers/unit area)? Can you suggest a biotic factor which must also be important? (See Haverty, LaFage and Nutting, 1974; LaFage, Haverty and Nutting, 1976.)

SELECTED REFERENCES

Beard, R.L. 1974. Termite biology and bait-block method of control. Conn. Agric. Exp. Sta., New Haven, Bull. 748, 19 pp.

Becker, G. 1969. Rearing of termites and testing methods used in the laboratory. In K. Krishna and F. Weesner (eds.), Biology of Termites, vol. 1. Academic Press, New York, pp. 351-385.

Haverty, M.I. and W.L. Nutting. 1975a. Natural wood preferences of desert termites. Ann. Entomol. Soc. Am. 68: 533-536.

Haverty, M.I. and W.L. Nutting. 1975b. A simulation of wood consumption by the subterranean termite, Heterotermes aureus (Snyder), in an Arizona desert grassland. Insectes Sociaux 22: 93-102.

Haverty, M.I., J.P. LaFage and W.L. Nutting. 1974. Seasonal activity and environmental control of foraging of the subterranean termite, Heterotermes aureus (Snyder), in a desert grassland. Life Sciences 15: 1091-1101.

Haverty, M.I., W.L. Nutting and J.P. LaFage. 1975. Density of colonies and spatial distribution of foraging territories of the desert subterranean termite, Heterotermes aureus (Snyder). Environ. Entomol. 4: 105-109.

Johnson, K.A. and W.G. Whitford. 1975. Foraging ecology and relative importance of subterranean termites in Chihuahuan Desert ecosystems. Environ. Entomol. 4: 66-70.

LaFage, J.P. and W.L. Nutting. 1978. Nutrient dynamics of termites. In M.V. Brian (ed.), Production Ecology of Ants and Termites, Cambridge University Press, London. pp. 165-232.

LaFage, J.P., M.I. Haverty and W.L. Nutting. 1976. Environmental factors correlated with the foraging behavior of a desert subterranean termite, Gnathamitermes perplexus (Banks) (Isoptera: Termitidae). Sociobiology 2: 155-169.

LaFage, J.P., W.L. Nutting and M.I. Haverty. 1973. Desert subterranean termites: A method for studying foraging behavior. Environ. Entomol. 2: 954-956.

Snyder, T.E. 1954. Order Isoptera, the termites of the United States and Canada. Tech. Bull. Nat. Pest Control. Assoc., New York. 64 pp.

Weesner, F.M. 1965. The Termites of the United States, A Handbook. Nat. Pest Control Assoc., Elizabeth, New Jersey. 70 pp.

12. Foraging Orientation of Honey Bees

Rudolf Jander
University of Kansas

Obtaining food is an item of overwhelming importance on the life agenda of all animals. The problem can be approached in many ways, but commonly involves some manner of active search requiring energy expenditure and time involvement. In cases where the food source is non-mobile, one would expect certain behavioral adaptations in orientation to minimize search time and distance and to increase foraging trip "profit".

One of the more thoroughly studied cases of foraging orientation among insects may be found in the honey bees, whose behavior has been experimentally approached from nearly every conceivable direction. To find enough nectar or pollen for a load, foraging bees usually have to visit a great many flowers over a wide area. It appears clear that honey bees, like bumble bees and many other Hymenoptera, increase their foraging efficiency by a variety of behavioral orientation mechanisms such as site and route fidelity when travelling from 1 clump of flowers to another.

But within a flower clump what sort of orientation occurs? Are the individual flowers visited in some particular order? Or randomly? What factors might be influencing orientation direction in the clump? Such questions are difficult to answer under natural field conditions because of the vast number of stimuli that the bees might be responding to. By using artificial flowers in a somewhat structured situation, we can control some of these stimuli and can experimentally examine some aspects of honey bee foraging orientation.

METHODS

Subject

Honey bees searching for nectar are highly suitable for the following experiments. They are available everywhere throughout the warm season, either from some local beekeeper or by mail order. A single standard hive box, containing 10 frames and a minimum of 2 to 3 pounds of bees with a queen, is fully adequate. The bees can be enticed to forage in almost any desirable location; it is even possible to have them fly through an open window into a closed room (such as the laboratory) where the experiment can be set up.

NOTE: Students hypersensitive to bee stings should not participate in this exercise.

Materials Needed

At least 20 flower models (Fig. 12-1; for construction tips, see Aids to the Instructor); repeating pipette, ejecting 2 µl per operation; 30% sugar water or 50% honey water; graph paper; electric fan (optional).

Procedure

The first part of this exercise has to be done several days in advance of the actual foraging orientation studies. (Your instructor may choose to set this part up for you.)

A. Training the Bees to the Testing Location

Choose a foraging location, preferably 20 to 50 m from the hive. (Short flight distances reduce the time necessary for training the bees to the testing site and increase the return frequency of the foraging bees.) Set up a row of flower models (20 or more), 1 every ½ m starting at the hive entrance. The bees will more easily find the first flower if they can walk directly onto it from the hive entrance. Fill each flower with approximately 20 µl of rewarding solution (sugar water or honey water). Once a group of bees has begun foraging on the row of flowers, keep the flowers constantly filled with reward. Simultaneously and gradually, take away the flowers nearest to the hive and place them at the other end as a continuation of the row. Thus the row of flowers is slowly "creeping"

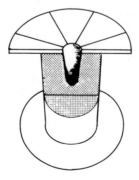

Fig. 12-1. Wooden flower model. The front upper half is seen cut away to expose the central boring. The diameters of the top disc and the bottom disc are 6 cm.

toward the prospective testing site. The speed of progression should be adjusted so that at least 5 bees are able to follow.

If the bees are finally required to fly through an open window into an enclosed room, the open window should be in the direction of the brightest light entering the room. Departing bees have a very strong habit of flying toward bright light, which is difficult to modify.

Shortly before the actual foraging orientation tests, capture all the foraging bees except 1, and put them into a cage. These will serve as stand-bys if the 1 selected forager should get lost.

B.   Observing and Testing Foraging Orientation

Arrange 6 flowers in a circle with a diameter of 30 cm (Fig. 12-2). Place a seventh flower in the center. Fill all the flowers with 2 μl of reward solution, or more if necessary to sustain reliable visitation rates. Refill as needed. Maintain a store of flowers containing reward, in order to quickly replace flowers that have been visited by the foraging bees.

While the bee is feeding in any 1 flower, place the flower (with the bee in it) into the center of the whole arrangement, where it replaces the flower which has been there before. Simultaneously, replace the flower visited by the bee previously with a reward-filled flower from the store of flowers. This procedure ensures that the foraging bee always starts from the center of the whole arrangement and then selects 1 of the 6 equally rewarding peripheral flowers. Thus, right in front of the observer, the bee performs on 1 spot a foraging flight pattern as if it were on a flower field without bounds.

While this goes on, the sequence of the flight directions of the foraging bee should be recorded. In clockwise order, number the 6 peripheral positions, which correspond to 6 directions of the compass. (It is convenient to have the sixth flower exactly north.) Record every foraging bout of the bee as a sequence of numbers between 1 and 6; each number represents a sector of the compass 60 degrees wide. Due to this recording procedure, the search path of the bee will be recorded as a polygonal path with constant lengths of the sides, and with angles at the vertices that are whole multiples of 60 degrees.

Repeat until so many foraging paths of the foraging bee are recorded that at least 100 flight directions between flowers have been accumulated.

C.   Analyzing Foraging Orientation for Directionality and Persistence

Count the number of times the 6 different flower directions have been visited. If the foraging path of the bee has been completely random, the frequencies with which the 6 flowers have been visited should be about equal. This null hypothesis can be tested by applying standard chi square analysis (see Appendix 1.1).

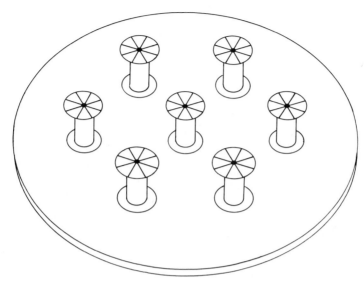

Fig. 12-2. Arrangement of 7 flower models as used in this exercise. The 6 peripheral "flowers" are equally spaced on a circle of 15 cm radius.

CAUTION: Randomness of orientation is a sufficient, but not a necessary explanation for a uniform distribution of the 6 flight directions. It is not difficult to imagine many different ways in which a foraging bee could systematically change its directions of flight. The data can then be inspected for any 1 such hypothesis of systematic orientational changes.

1) If your data appear to show directionality: Should the hypothesis of a uniform distribution of directions be refuted, then it is most likely that the bees preferred 1 particular direction of the compass. Prepare a frequency histogram of your data; directionality should become obvious from a visual inspection.

2) If your data show a uniform distribution of directions: Persistence is the tendency of a foraging bee to leave a flower in the direction opposite from which it was approached. In order to analyze your data for this phenomenon, you must determine conditional frequencies by means of a transition frequency matrix. Plot 6 frequency histograms, 1 for each approach direction. For a particular approach direction histogram, enter the number of times the bee left in each direction. The condition for entering a data point of a particular departure direction into 1 of these histograms is the previous approach direction prior to departure. (For example, if the bee visited the flowers 2, 3, 3, 6 in that sequence, you should have 3 entries, one (3) into the "approach direction # 2" histogram and two (3,6) into the "approach direction # 3" histogram.)

Direct visual inspection of these 6 conditional frequency histograms should reveal whether your bee exhibited persistence. For instance, when the bee approached the sixth flower, persistence would mean that the sixth flower and its neighbors (flowers 5 and 1) would be more likely than the other 3 flowers to be approached next.

SUGGESTIONS FOR FURTHER STUDY

1. Wind Direction and Foraging Orientation

The single most important factor that may induce a bee to prefer a particular direction of flight is wind. Therefore, natural wind direction should be recorded during the experiment if it is performed in an unsheltered area outside. If your experiment was performed in a sheltered place, you may wish to repeat it while air flow is artificially introduced with an electric fan. Wind speeds of 2 m/sec have noticeable effects on the directional preference of the foraging bee.

2. Further Statistical Analyses of Foraging Orientation

You may wish to use your data for various statistical exercises leading to a deeper understanding of the bee's foraging strategy. Two crucial statistical parameters to extract from the data are overall mean direction, if any, and mean conditional departure direction.
For exact evaluation of either of these, consult a book on biometry which contains a chapter on circular statistics (such as Zar, 1974).

SUMMARY QUESTIONS

1. Given the experimental outcome, discuss possible underlying mechanisms and adaptive significance of randomness, upwind orientation, and persistence in foraging.

2. Honey bees are capable of sun-compass orientation (v. Frisch, 1967). How, in this experiment, would you test whether the forager uses the sun as a directional cue for controlling the direction of flight between flowers?

3. Assume that during 1 visit the "perfect bee" harvests any 1 flower only once and simultaneously minimizes its cumulative flying time (or distance). Invent an optimal foraging path within a patch of flowers. Compare your solution to that of the bee in your experiment, and interpret the difference.
SELECTED REFERENCES

Frisch, K. v. 1967. The Dance Language and Orientation of Bees. Bees. Belknap Press, Harvard University Press, Cambridge, Mass.
Waddington, K. D. 1980. Flight patterns of foraging bees relative to density of artificial flowers. Oecologia 44: 199-204.
Zar, J. H. 1974. Biostatistical Analysis. Prentice Hall, Englewood Cliffs, New Jersey.

FOOD RECOGNITION, ACCEPTANCE,
AND REGULATION

13. Host Acceptance by Aphids: Effects of Sinigrin

Lowell R. Nault
Ohio State University

On what specific basis does an insect recognize that an item is food? For plant feeders -- a category that includes about half of all the Insecta -- it is clear that plant odor and taste mediate the first steps in feeding.

Because much of the support enjoyed by entomologists derives from the economic importance of insect feeding, many detailed studies have been directed toward discovering the nature of those aspects of plant chemistry involved in insect feeding. Many diverse chemicals not known to have importance in plant growth or metabolism (and for that reason commonly called secondary substances) have been identified. These have a variety of interesting effects on insect behavior; the bulk of the evidence has suggested that the chemicals' primary significance is to deter feeding. On the other hand, the various plants used as food by a particular insect species, genus, or even family often share similar secondary substances although they may differ widely in other respects. This suggests that insects have often been successful in turning the tables, using the plants' chemical defenses as distinctive cues to single them out.

Plants in the Cruciferae, or mustard family, include a wide variety of commercially important vegetable crops -- radish, mustard, turnip, watercress, cabbage, cauliflower, broccoli, brussels sprouts, etc. -- characterized by a sharp, strong odor and/or flavor caused by certain mustard oils and their glucosides. The Cruciferae have a characteristic insect fauna which includes a number of Lepidoptera, Diptera, Coleoptera, and Homoptera. The host range of these insect species is restricted to the Cruciferae and a few closely related plant families containing mustard oils and their glucosides. For a number of these insects, these chemicals stimulate feeding and oviposition (Whittaker and Feeny, 1971).

One such chemical which has been relatively well studied is the mustard oil glucoside, sinigrin. It has a number of effects, which differ with the life stage and species of feeding insect. Among aphid species which are not Cruciferae feeders, sinigrin acts as a feeding deterrant (Nault and Styer, 1972). However, it has stimulatory effects on the feeding behavior of the cabbage aphid and turnip aphid (Wensler, 1964; Moon, 1967; Nault and Styer, 1972). Sinigrin stimulates stylet penetration into the phloem of the plant,

and assists aphids in locating the phloem sieve elements, the primary
aphid feeding site.

Most phytophagous insects accept only a limited range of foods.
If only a single food plant is accepted, the condition is known as
monophagy. If several related food plants are preferred, it is
called oligophagy. Other insects will accept a wide variety of foods
(polyphagy), although they may still show decided preferences. A
restricted diet may not mean that unacceptable foods are
non-nutritious. Cabbage and turnip aphids will not normally feed on
such plants as broadbean, pea, clover, cucumber, squash, potato,
corn, or barley. However, if cut leaves of these "non-hosts" are
placed in a 1% water solution of sinigrin, the resulting food is not
only acceptable but quite adequate. Broadbean, for example, has been
experimentally converted into a suitable host for the turnip aphid;
18 consecutive generations have been reared on treated leaves (Nault
and Styer, 1972).

In this exercise, you will have the opportunity to test the
effects of sinigrin on feeding by cabbage, turnip, and pea aphids.
You will become familiar with "choice test" and "no choice test"
procedures for studying animal feeding, examine concentration
effects, and develop dose response curves.

METHODS

Subject

Cabbage and turnip aphids are oligophagous, feeding entirely on
plants within the family Cruciferae. Both species are pests of
cabbage, turnip, radish, and other cole crops, and may be
field-collected from these plants. With the exception of the broadly
polyphagous green peach aphid, Myzus persicae (Sulzer), no other
aphid species will feed on the Cruciferae containing mustard oil.
Conversely, the pea aphid, Acyrthosiphon pisum (Harris), is
oligophagous, feeding on plants in the family Leguminoseae; hosts
include pea, broadbean, alfalfa, and clover. (If pea aphids are not
available, other species which are not Cruciferae feeders may be
substituted.)

The cabbage aphid, Brevicorne brassicae (L.), is also available
from stock culture and may be reared in the laboratory (Nault, 1969)
on turnip, cabbage, or kale. The turnip aphid, Lipapis erysimi
(Kaltenbach), also called Hyadaphis erysimi or Aphis pseudobrassicae,
may be similarly reared on turnips or radish. Either or both may be
used in this exercise.

See Aids to the Instructor for help identifying these species.

Materials Needed

Aphid cultures; leaves from cruciferous and non-cruciferous
plants (turnip, radish, broadbean, and potato are recommended);
sinigrin; petri dishes (100 x 20 mm); single edged razor blade or
scissors; ¼ dram vials; non-absorbent cotton; fine camel's hair
brushes; felt tip markers.

Procedure

A. Choice Tests for Sinigrin Effects on Host Selection by Cruci-
ferae-feeding Aphids (Fig. 13-1).

    Obtain 8 vials with non-absorbent cotton plugs, 4 petri dishes
with lids, a single edged razor blade, and a fine brush. Fill 2 of
the vials 2/3 full with a 1% aqueous solution of sinigrin. Fill the
other 6 to the same level with water.
    Use the razor blade to cut the petiole of 3 leaves each of pota-
to and broadbean, and the petiole of 2 turnip leaves. (Remember that
the leaves must be small enough to fit within the petri dishes.)
Place the leaf petioles in the vials as indicated below, plugging the
opening of each vial with non-absorbent cotton so that when the vials
are placed on their sides the fluids will not spill. Set up your
petri dishes to contain 4 choice tests (labelled!) as indicated on
the sample data sheet (Table 13-1).
    Using the fine brush, gently place 10 adult turnip or cabbage
aphids in each dish. Replace lids, and place dishes in the location
specified by your instructor. Keep all the petri dishes at about
70°F for 24 hours. Then count the number of adult and first instar
aphids on each leaf. Be careful while counting aphids; they are

TABLE 13-1. Sample Data Sheet for Choice Tests of Aphid Feeding.

Aphid species:				Investigator(s):				
				Total # Aphids		Total Larvae	Percent Adults	
Compar-ison	Treat-ment	Leaf Treated	Soln.	On Leaf	Off Leaf	per Adult	on Leaf	
				L A	L A*			
choice test 1	1 2	broadbean broadbean	H_2O sini-grin					
choice test 2	1 2	potato potato	H_2O sini-grin					
(control)	1 2	broadbean turnip	H_2O H_2O					
(control)	1 2	potato turnip	H_2O H_2O					

* L = larvae, A = adults.

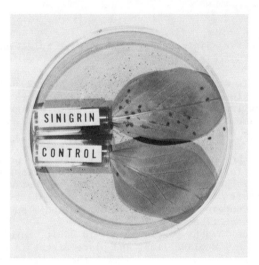

Fig. 13-1. Turnip aphid adults and larvae on a broadbean leaf treated with a 1% aqueous solution of sinigrin. Aphids settled on the undersurface of the leaf, which was turned over for this photograph.)

Fig. 13-2. Pea aphid adults and first instar larvae on water-treated broadbean leaf. (Aphids settled on the undersurface of the leaf, which was turned over for this photograph.) A single adult has settled on the leaf treated with a 1% aqueous solution of sinigrin.

easily disturbed. Most will be on the undersurfaces of leaves, so vials must be turned or picked up to make the counts. Also count aphids which are not settled on leaves.

The first instar aphids are of particular interest. Viviparous, parthenogenetic aphids such as those used in this exercise normally deposit larvae on the undersides of leaves while they are feeding. These larvae rarely wander from leaves on which they were deposited, and adults seldom larvaposit unless feeding. Therefore, counting these first instar aphid larvae provides a measure of feeding activity superior to that of counting only adults, which may wander about the dish during an experiment.

B. Effect of Sinigrin on Non-cruciferous Feeders

Following the procedure outlined above, provide a petri dish with a pair of broadbean leaves, 1 treated with 1% sinigrin solution and the other with water. (Label both!) Introduce 10 adult pea aphids. After 24 hrs, record location and larvaposition data as before.

C. No Choice Tests for Sinigrin Concentration Effects on Host Suitability

Obtain 4 vials, label, and fill with aqueous sinigrin in concentrations of 10, 10^2, 10^3, and 10^4 parts per million. (Which of these represents the concentration used in the activities above?) Fill a fifth vial with water only.

Place 1 broadbean leaf in each vial, plug with cotton, and place in a petri dish. Add 10 adult pea aphids to each dish and replace cover. After 24 hrs, observe and record data as before.

Replicate using either the turnip aphid or the cabbage aphid.

D. Data Analysis

Pool your data with other students to obtain an appropriate number of replications. The extent of your statistical analysis will be specified by your instructor. As a minimum, your laboratory write-up for part A should include the computations of results listed in Table 13-1. For part B, performing a chi square test (see Appendix 1.1) on your data is suggested. For part C, develop dose response curves for the pea aphid and for the turnip or cabbage aphid. Average the larvaposition responses of all the individuals of the species for each concentration, including the control. Divide the control average by the experimental concentration average to get a value of average larvaposition as a percent of control. Plot each of these larvaposition values on a graph, and connect the points to get a dose response curve for the aphid species.

SUGGESTIONS FOR FURTHER STUDY

If you have time, you may wish to rear 1 generation of turnip aphids on sinigrin-treated broadbean leaves, and compare these to those reared on the turnip host. This takes about 8 days. Is there a difference in mortality or size of resultant adults?

SUMMARY QUESTIONS

1. Were the sinigrin-treated broadbean and potato leaves as attractive to the turnip or cabbage aphids as were the turnip leaves?

2. At what dosage did sinigrin have a significant negative effect on the settling and larvaposition of the pea aphid? At what dosage did sinigrin have a significant positive effect on the settling and larvaposition of turnip or cabbage aphids on broadbean leaves?

SELECTED REFERENCES

Moon, M.S. 1967. Phagostimulation of a monophagous aphid. Oikos 18: 96-101.

Nault, L.R. 1969. Laboratory rearing of aphids. J. Econ. Entomol. 62: 261-262.

Nault, L.R. and W.E. Styer. 1972. Effects of sinigrin on host selection by aphids. Entomol. Exp. Appl. 15: 423-437.

Wensler, R.J.D. 1962. Mode of host selection by an aphid. Nature, London 195: 830-831.

Whittaker, R.H. and P.P. Feeny. 1971. Allelochemics: Chemical interactions between species. Science 171: 757-770.

14. Blow Fly Feeding Behavior

John G. Stoffolano, Jr.
University of Massachusetts

Faced with "appropriate" food at the "right" time, most insects will feed to a point and then stop. What causes an insect to begin feeding? What finally causes it to stop? Such questions have received their most detailed investigation in the cause of the blow fly, Phormia regina, primarily through the work or Dethier (1962, 1976) and his associates. Of all the animals, the only one that rivals the blow fly when it comes to understanding the mechanisms regulating feeding is the laboratory rat. Even then, the insect system is better understood.

This exercise is designed to acquaint you with the experimental animal, the fly, and some of the techniques that have helped us understand the mechanisms regulating feeding in this insect. You will discover through experimentation how adult flies taste their food, what sugar concentrations a blow fly will respond to as food, and what role the nervous system plays in feeding regulation. (If time permits, you may also conduct experiments to determine some of the factors influencing feeding behavior in blow flies of different ages.) You will learn how to behaviorally determine feeding acceptance thesholds and how to carry out an operation in an insect that will in turn influence behavior.

METHODS

Subject

Blow flies (Calliphoridae) and flesh flies (Sarcophagidae) are the common "house flies" of the western parts of the United States, especially the Southwest. Often metallic green or blue, most are scavengers, the larvae living in carrion and similar materials. They are, however, easily reared under laboratory conditions. Both larvae and pupae may be obtained at very nominal cost from biological supply companies or captured in the field.

Although Phormia regina has been the species traditionally used, interesting results also may be obtained from other common flies (such as Sarcophaga, Musca, etc.).

Materials Needed

Adult blow flies, some starved 1-2 days but given water, others not starved; distilled water; 7 solutions of glucose or sucrose (conc. .005M, .01M, .05M, .125M, .25M, .5M, and 1M); blue food coloring; liver juice; applicator sticks with Tackiwax balls on the end; stick holders; CO_2 tank; forceps; petri dishes (8 per team); containers of crushed ice; dissecting microscopes; insect pins; balance; minuten pins; grease pencils.

Procedure

A. Anesthetizing and Preparing Blow Flies for Feeding Experiments

Working in pairs, obtain a container with at least 10 flies in it. Using the CO_2 tank as demonstrated by the laboratory instructor, anesthetize your flies and then place their container on top of the crushed ice. What will the ice do? If CO_2 is not available, flies can be immobilized by placing the container into the refrigerator or into the ice until they become inactive.

As demonstrated, embed flies by their wings into the Tackiwax at the end of the applicator sticks. Put 2 flies per stick, back to back (see Fig. 14-1). Before embedding the wings into the Tackiwax, warm the wax end by enclosing it in your palm. This will soften the wax, making it easier for you to push the wings into the wax. Both wings MUST be firmly embedded so that the thorax is firmly pressed against the wax, or the flies will work their wings loose. Place the sticks upright in the holders. Let the flies remain on the sticks for about 20 min. before testing. Why?

While you are waiting, obtain 8 plastic petri dishes. Label the inside bottoms with a grease pencil before adding solutions to them. In 1 dish, add about 5-6 mm distilled water. In the others, add the 7 different concentrations of sugar to the same depth. Also complete part B of this exercise.

B. Determining the Location of the Sense of Taste

Obtain 1 of the already prepared flies from the instructor. To determine where other areas for the sense of taste are located on the fly (in addition to feet and proboscis), take the 1M solution of sugar and dip the end of an applicator stick into it so that a small drop adheres to the end. Touch the stick to various areas of the fly to see what happens. Record the locations touched and the type of response.

Place the same fly under the dissecting microscope. Notice all the hair-like structures covering the body, feet, and proboscis. Closely examine the proboscis under high magnification. If time permits, remove the proboscis and place it between a cover slip and microscope slide for examination under a compound microscope. Fig. 14-2 is a diagram illustrating the general structure of an individual labellar chemosensillum in Phormia. Notice that there are usually

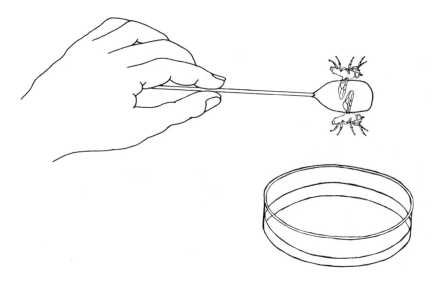

Fig. 14-1. Student preparing to touch the tarsal chemosensilla of a fly to a stimulating sugar solution. Note the back-to-back position of the flies and how their wings are embedded into the Tackiwax ball at the end of the applicator stick.

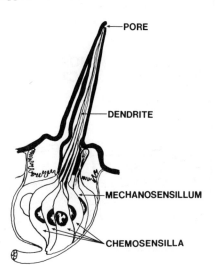

Fig. 14-2. Schematic diagram of a labellar chemosensory hair of a fly. Note that the dendrites of the 4 chemosensilla pass through the hair shaft to the pore while that of the mechanosensillum terminates at the base of the hair.

5 bipolar neurons at the base of each chemosensory hair, 4 being chemosensilla and 1 a mechanosensillum. The dendrites extend from the cell bodies up through the hair shaft to the tip where they are exposed at the pore to the external environment. What do the taste papillae on your tongue look like? Do they compare in any way with the chemosensilla of the fly's "tongue"?

At this point, check with the instructor as to the results from this section.

C. Determining the Behavioral Tarsal Acceptance Theshold

When a blow fly confronts a sugar solution, it undergoes a several stage stimulus-response sequence which eventually leads to a decision whether or not to begin feeding. First, the fly steps into the liquid and detects the sugar as it contacts chemosensory hairs on the tarsi. If the stimulus is strong enough, the fly extends its proboscis (see Fig. 14-3), thus bringing the labellar hairs in contact with the food. If stimulation of the labellar chemosensilla is sufficient, the labellar lobes open and labellar interpseudotracheal papillae are stimulated. The chemosensilla of these 3 different areas may have different thresholds, and only after the papillae are stimulated does the fly begin ingesting the solution.

Basically, there are 2 types of thresholds involved in feeding: receptor potential threshold (r.p.t.) and behavioral threshold. The r.p.t. is determined using sensitive electronic equipment to monitor the response of individual neurons to various compounds. The behavioral threshold measures an overt response (acceptance threshold)

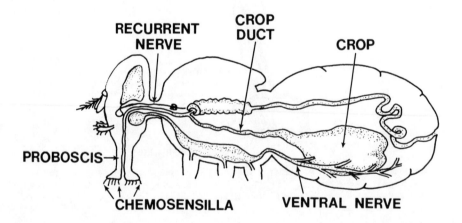

Fig. 14-3. Diagram of a longitudinal section of a fly showing the arrangement of the recurrent nerve, ventral nerve, crop duct and crop. Note the proboscis and the hair-like chemosensilla at the end.

of the whole animal or failure of response (rejection threshold). The behavioral tarsal acceptance threshold is defined as the concentration of sugar solution to which 50% of the individuals tested responded by extending their proboscis. (Remember that it is possible to initiate a behavioral response to a stimulating sugar by only contacting 1 labellar hair. This situation, however, seldom if ever occurs in nature.)

Before testing the flies, determine your own taste threshold to the 7 glucose or sucrose solutions you have before you, starting with the lowest and going to the highest. Score a positive response as +, and negative as -. Then, using the procedure as discussed with the instructor, test 10 starved flies with the solutions, again starting with the lowest concentration. Prior to testing these solutions, permit each fly to drink all the distilled water it will imbibe. Why?

The data for this section can be handled by pooling the class results and presenting them in the form of a histogram with the vertical axis being the percentage responding and the horizontal axis being the 7 different concentrations of sugar being tested. (What are other ways of treating the data?) Now test the sugar solutions using the fly.

If time permits, let each fly drink all the 0.5M sugar solution it will take. Wait 10 min., then retest all the flies as above. Has there been a change in threshold? If so, how can you account for this?

D. Removal of Sources of Negative Feedback Governing Blow Fly
    Feeding

Feeding stops when negative feedback from internal receptors inhibit brain input from the external chemosensilla that elicit feeding. This negative feedback is a response to a number of variables, including concentration of food substance and/or crop capacity, level of sensory stimulation, activity level, and reproductive state (Stoffolano, 1974). By carrying out an operation, one can eliminate this feedback. Alternatively, one can circumvent the feedback loop by stimulating the chemosensilla on the labellum rather than the tarsi (Nelson, 1972).

To remove sources of feedback by cutting the recurrent nerve or ventral nerve (see Fig. 14-3), obtain 10 adult blow flies on individual sticks attached to Tackiwax. Identify each fly by placing a number below the hole on the stick holder. Following the technique demonstrated by the laboratory instructor, operate and remove (break) the ventral nerve of 5 flies.* These flies are termed the treated flies. Cut the cuticle of the other 5 flies in the same place as

---

*A film loop, "Feeding Behavior of the Blowfly" by Vincent Dethier, is available from Harper and Row (Cat. #04-78198, Library of Congress Card No. 77-706630) and shows the technique for cutting the recurrent nerve. It is easier, however, to cut the ventral nerve.

that of the treated flies, but do not cut or injure the ventral nerve. These flies will serve as the sham controls.

Following the operation, record the weight of each fly (including stick and wax). In experiments involving the measurement of liquid intake by weighing, it is easier to weigh the flies in groups (treated vs. sham controls). This should save time, and will get around the problem of sensitivity of the different types of balances available.

After a 10 min. adaptation period, permit the flies to take 1 uninterrupted drink of 1M sucrose. Weigh the flies again. Calculate the weight increase for the controls (individually or as a group) versus the treated flies. Pool the class data for examination and analysis. What kind of statistical test could one run on these data?

SUGGESTIONS FOR FURTHER STUDY

1. The Influence of Age Upon Feeding Behavior

When a blow fly first emerges from its puparium, it expands itself by pumping air, making pharyngeal movements like those used later in feeding. Additionally, very starved flies will extend their proboscis -- the first act of feeding -- without any chemical stimulation, merely in response to mechanical stimulation. How are age and physiological stage related fo fly feeding? When an adult does not feed, is it because something is inhibiting its feeding, or is the neural circuitry not "hooked up"? The majority of research on factors influencing failure to feed in blow flies has been done on flies that have been starved and then fed. Newly emerged flies, however, have not been fed, and failure to feed must involve other factors.

Carefully remove adults from the puparia at various stages of development. Using various testing solutions of sucrose, try to get them to extend their probiscis. Also attempt to get newly emerged flies to respond to a sucrose solution.

2. Food Diversion in the Adult Blow Fly

Most flies have their digestive tract partitioned so that food can be diverted for storage to a sac-like structure called the crop. No digestion occurs here but instead takes place in the midgut. Take 2 of the flies on sticks, and let them drink all of the sugar water (0.5M blue food colored solution) they want. Using 2 more, do the same thing with liver juice. Submerge them under water in a petri dish, and carefully dissect them. The easiest way to complete the dissection is to carefully grab both sides of the abdomen with the forceps. As you pull the cuticle will pull away, leaving the digestive tract exposed. Once this has been done, carefully remove other pieces of cuticle. Locate the crop, midgut and hindgut (Fig. 14-3). Notice where each type of food (carbohydrate and protein) is directed in the digestive tract.

FINAL REMARKS

Ultimately, an understanding of the feeding behavior of an insect has some practical application. Areas where this basic research has been applied include:

- Development and utilization of attractants, repellents, and deterrents.
- Plant breeding programs for insect resistance.
- Analysis of arthropod-borne diseases.
- Development of baits.
- Development of rearing techniques for laboratory colonization.

Further information on measuring feeding thresholds may be obtained by reading Thompson (1977). Barton Browne (1975) reviews the literature on regulatory mechanisms.

SUMMARY QUESTIONS

1. What is the chain of events (stimuli and responses) that ultimately leads to ingestion of a food substance by the blow fly? Will these events be the same if a fly encounters a solid or semi-solid source of food?

2. Can you think of some insects where initiation of feeding does not start with the tarsi contacting the food source? In these instances, what initiates the feeding response?

3. Many hair-like structures on insects are in fact not chemosensilla. How might one determine whether particular setae are chemosensilla or not?

4. Briefly discuss the results of the operation experiment to remove sources of negative feedback. What does the term "cybernetics" mean? How does it relate to this problem of feedback?

5. What might some of the factors be that would influence feeding in newly emerged (teneral) flies?

6. Discuss the advantages and disadvantages of food diversion in the digestive tract of an adult fly.

7. Define the terms monophagous, oligophagous, polyphagous, phytophagous, and hematophagous.

SELECTED REFERENCES

Barton Browne, L. 1975. Regulatory mechanisms in insect feeding. In J.E. Treherne, M.J. Berridge and V.B. Wigglesworth (eds.), Advances in Insect Physiology, vol. 2. Academic Press, New York, pp. 1-116.

Dethier, V.G. 1962. _To Know A Fly_. Holden-Day, San Francisco.

Dethier, V.G. 1976. _The Hungry Fly_. Harvard University Press, Cambridge.

Keeton, W.T., M.W. Dabney, and R.E. Zollinhofer. 1968. Chemoreception and behavior in the blowfly. In _Laboratory Guide for Biological Science_. W.W. Norton and Co., Inc., New York, pp. 167-176.

Nelson, M. 1972. Negative feedback control in the blowfly -- a laboratory exercise. The Physiology Teacher 1: 1-3.

Ross, J.A. 1975. Feeding behavior in the blowfly. In E.O. Price and A.W. Stokes (eds.), _Animal Behavior in Laboratory and Field_, 2nd ed. W.H. Freeman, San Francisco, pp. 27-29.

Stoffolano, J.G., Jr. 1974. Control of feeding and drinking in diapausing insects. In L. Barton Browne (ed.), _Experimental Analysis of Insect Behaviour_. Springer-Verlag, New York, pp. 32-47.

Thompson, A.J. 1977. Measurement of feeding threshold in the blowfly _Phormia regina_ Meigen. Can. J. Zool. 55: 1942-1947.

15. Communication in Mealworm Beetles

Ronald L. Rutowski
Arizona State University

By definition, communication, the exchange of information between 2 animals, has occurred when the behavior of 1 animal (the sender) alters the behavior of the other animal (the receiver). Such a biological definition usually further stipulates that the receiver's response be reproductively advantageous to both sender and receiver; this distinguishes communicatory phenomena from cases of prey detection by predators (and vice versa), aggressive mimicry, and other effects of animal signals that are thought to be incidental to their evolved function.

Studies of animal communication follow a certain logical order. First, one determines a way to tell whether an animal has actually received a signal. The specific cues being sent are often not obvious or apparent to a human observer, so one needs a well-defined behavioral assay -- a response that permits assessment of whether or not an animal has received the information.

Second, using the bioassay one learns which sensory modalities are actually involved in the communication. The most general types of signals used in sexual discrimination by insects, for example, are chemical, visual, and mechanical (auditory and tactile). Although insects may also be sensitive to magnetic and electrical fields, it appears doubtful that these play a role in communication.

Having experimentally established that the important information in the signal is carried in 1 of these modalities, one can then examine specific aspects of the signals and, through further experiments, determine the specific features of the sender's signal to which the receiver is responding. For example, what specific chemicals are involved, or what wave lengths of light are important? At all stages in this deductive process, hypotheses are tested with the bioassay.

Defining the complete set of environmental stimuli that elicit a response is a potentially very complex task. The study of communication may be greatly simplified, however, by asking not for the complete spectrum of ways conspecific females, for example, are recognized, but rather how males distinguish females from other males. Thus, the problem becomes one of determining the actual differences between 2 seemingly identical stimuli.

In this exercise we will examine the sexual signals given off by female mealworm beetles that permit males to discriminate the females from males. Our goals will be to formulate a bioassay and experimentally to determine the signal modality involved in the male's discriminations, in the process becoming acquainted with the logic of experimental studies of animal communication.

METHODS

Subject

Tenebrio are small (13-17 mm long) black or dark brown beetles which feed upon stored grain in both larval and adult stages. Larvae are often sold as fish bait and pet food, and are easily maintained in laboratory culture.

Determining the sex of mealworms is not easy, but is done by gently squeezing the abdomen of an adult until the genitalia evert (Fig. 15-1). Males have a small needle-like aedeagus (penis) that will protrude from between the dorsal and ventral sclerites of the genital apparatus; their ventral sclerite is notched. Females have a tube-like ovipositor several mm in length.

Materials Needed

Mature mealworms of both sexes; dissecting microscope for sexing mealworms; petri dishes (or other glass covers); stopwatches or watches with second hand; quick-drying enamel or acrylic paint; brushes; ethanol; glass rods; data sheets.

Procedure

A. Observation of Interactions

Your instructor will have sexed a quantity of adult mealworms and isolated males and females for several days before this exercise to increase their responsiveness to the opposite sex during the class period. Obtain several beetles of both sexes, and give each a sex-specific mark of paint on its pronotum.

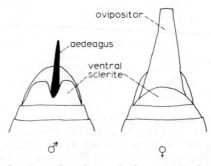

Fig. 15-1. Ventral view sketches of the terminal abdominal segments of the male and female mealworm. In both cases the abdomen was squeezed gently but firmly until the aedeagus or ovipositor was extruded.

Place several males and females together on a paper towel, and cover them with an inverted glass petri dish to restrict their movements and encourage contact. It may be helpful to dim the room lights at this point. Observe and make written notes of all interactions. Record the sex of both participants, the form of the interaction, and its outcome. Pay special attention to the behavior of males and females in interactions that lead to copulation. You may want to limit the number of individuals under the petri dish after a time to minimize interference in male-female interactions. With a stopwatch, time the duration of male-male and male-female interactions. At the end of this part of the exercise you should write out clear descriptions of the interactions males have with conspecific mealworms of both sexes. With your lab mates discuss the features of the male's behavior that would permit you to assess whether a male is responding to a given stimulus as a male or a female.

B. Experimental Analysis of Sex Discrimination

To clearly define the circumstances under which we can say that a given animal or subject is treated as a female by males, we need to be able to answer at least some, if not all, of the following questions.

1) What specific male responses will we require the male to perform to show that he is treating a stimulus as a female?

2) How long must the male be in the proximity of a stimulus before we can say with certainty that he has not responded to it?

3) If the male does not respond, how can we decide whether it is a function of the stimulus or of the internal state of the male?

These points should be discussed with the entire class, and a consensus reached so that all students will be collecting data in the same way. Data for the entire class can then be pooled.

Next, discuss the stimuli you need to present to determine the signal modality used by males for sexual discrimination. Usually, this series of experiments takes on the form of a process of elimination. (Why?) For example, you may wish to present males with freeze-killed males and females, males and females that have been killed and air-dried for a week or so, and as controls, live males and females.

Your instructor placed 30 female mealworm beetles in 10 ml of absolute ethanol no more than 24 hrs. before the laboratory period. Dip a small glass rod (about the diameter of a mealworm adult) into the female-filled alcohol for a few sec., air-dry the glass rod, and present it to males. Do not forget to run control experiments with glass rods dipped in clean ethanol. Do the males respond differently to the 2 stimuli? If so, in what way? How would you predict males should respond to the bodies of females that had been washed twice in ethanol and then air-dried before presentation? What do the responses to the glass rods tell you about the role of visual signals in mealworm courtship?

SUMMARY QUESTIONS

1. Is the chemical alone sufficient to elicit the full range of courtship behaviors from the males? If not, what roles do the other stimuli play? How might you experimentally document their roles?

2. Over what distance does the signal act? How might this be adaptive, given the natural history of mealworm beetles?

3. Are all your results in full agreement with the conclusion that males use chemical signals for sexual discrimination? If not, why not? Is the bioassay without fault?

4. What may be the ultimate reasons for the fact that mealworms use chemical, rather than visual or mechanical, signals for sexual discrimination? (Again give careful consideration to the natural history of mealworms.)

5. What is the nature of the chemical involved in the communication system? What are its structure, physical properties, and source? How would you experimentally answer such questions?

SELECTED REFERENCES

Birch, M. C. 1978. Chemical communication in pine bark beetles. Amer. Sci. 66: 409-419.

Rudinsky, J.A. and R.R. Michael. 1972. Sound production in Scolytidae: Chemostimulus of sonic signals by the Douglas-fir beetle. Science 175: 1386-1390.

Shorey, H. H. 1976. Animal Communication by Pheromones. Academic Press, New York.

Tschinkel, W., C. Willson and H.A. Bern. 1962. Sex pheromone of the mealworm beetle (Tenebrio molitor). J. Exp. Zool. 164: 81-86.

Tumlinson, J.H., M.G. Klein, R.E. Doolite, T.L. Ladd and A.T. Proveaux. 1977. Identification of the female Japanese beetle sex pherome: Inhibition of male response by an enantiomer. Science 197: 789-792.

Wilson, E.O. and W.H. Bossert. 1963. Chemical communication among animals. In G. Pincus (ed.), Recent Progress in Hormone Research, Vol. 19, Academic Press, New York, pp. 673-716.

16. Chemical Recruitment in the Fire Ant

Janice R. Matthews
University of Georgia

Ants communicate food source location by laying chemical trails. The trails also function to recruit additional colony members to the the food, thus maximizing food exploitation. The objectives of this exercise are to study trail laying behavior and recruitment in the fire ant, Solenopsis, under laboratory conditions.

METHODS

Subject

Originally introduced into this country from South America, the imported fire ant, Solenopsis invicta Buren, is an important pest in the southern United States from the Carolinas to Texas. Solenopsis richteri Buren is a closely related red form with a much more restricted range in Mississippi and Alabama. Both species have been used successfully in this exercise, and colonies can be maintained in the laboratory for long periods. Collection and care are outlined in Aids to the Instructor.

Materials needed

Ant colonies in pails; foraging arena set-up (Fig. 16-1); food items (mealworms, Drosophila, crickets, honey); stop watches; forceps; cover slips; meter stick; hand-held counters; disposable pipettes.

CAUTIONS:

1. Fire ants have a painful sting, with severe after effects in some individuals. With a little care, no one in the laboratory should have to learn of this first-hand! Since fire ants are well deserving of their name, treat them with due regard.

2. Take special care that nothing touches against the glass platform or bottles supporting the feeding arena; carelessness can result in a bridge over which the entire colony may escape.

3. Do not bump the ants or counter. Ants are sensitive to substrate vibrations.

105

4. Do not absent-mindedly lean against arenas or colonies, or ants will crawl off onto you.

Procedure

A. Behavior of Foraging Ants

A few hours prior to the laboratory session the instructor will have set up ant colonies and adjacent foraging arenas as in Fig. 16-1 (see also Aids for the Instructor). An hour or so before class the bridges connecting ant colonies to the arena should have been connected in order to allow time for a number of ants to find their way to the arena and begin exploring it.

How do individual ants react to different types of food? How do ants behave toward other foragers after discovery of a new food source?

Follow an ant until it is at least 15 cm from the bridge attachment. Note its path and any reaction upon meeting other workers. Place a freshly killed fruit fly next to the ant, and record the ant's subsequent behavior. Especially note any use made of the antennae, mouthparts, legs, or abdomen. If the ant encounters other foragers, what happens? Repeat this procedure twice more, using first a drop of honey on a cover slip, then a freshly killed cricket or mealworm. Describe any differences you observe using a data sheet like that in Table 16-1.

Next attempt to observe the initial discovery of a food source and the recruitment of foragers. Place a freshly killed cricket or mealworm on a cover slip on the foraging table near the bridge. Be patient and be ready to note the time of initial discovery. Also, check ahead for subsequent instructions under part C.

Carefully observe a forager before and after it discovers the

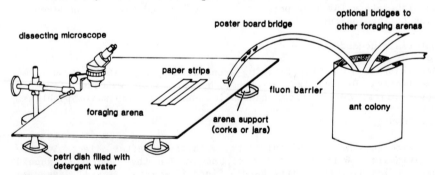

Fig. 16-1. Laboratory colony of fire ants in a bucket, with overpasses leading to the foraging arenas where observations are made using the dissecting microscope. Bridges are made from 3 cm wide strips of posterboard taped to the foraging arena and arched into the center of the bucket with care that the rim is not contacted. The arena may be of any convenient size and is supported on bottles or corks set in the center of detergent-filled petri dishes to serve as barriers to contain the ants.

bait. Describe how its behavior changes following discovery of food, and especially how it acts upon encountering other workers as it tries to return to the nest the first time after bait discovery. If it lays a trail, try to observe exactly how the trail is laid.

Suggestion: If after 15 min or so, a forager has not discovered the bait, there are 2 ways you can "cheat": first, present the bait directly to a forager; second, if artificial trail substance (made from ant abdomens extracted in pentane or ether) is available, streak an artificial trail, part way to the bait as demonstrated by your instructor.

B. The Information Communicated by a Trail

Do the ants really "need" the trail once food is found? Is food direction "told" by the trail?

Active trails should be visible crossing the strips of paper on the foraging arena connecting with trails from the overpass to the food. Label the 3 papers 1, 2, and 3 and for each of the following manipulations record the number or ants on each paper 30 sec and 1 min after the start of the experiment. Carefully move 1 of the papers over which the trail passes, sliding it temporarily out of line with the others. Does this appear to confuse the ants? After a min or so, return the paper to its original position. What happens? Next, gently rotate one of the end strips of paper 90°, so that is is at right angles to its original position. What happens?

Does the trail carry any information about direction? (In other words, can the trail supply information telling the ant at which end the food is?) Reverse the center strip of paper 180° from its original position, being careful to have the 2 ends of the trail connect again. How do the ants react?

What other cues might the foragers use to orient themselves to the nest or food source? How might they be tested for?

C. Control of Foraging Behavior

What happens to the trail when the food disappears? One explanation for the seemingly complex foraging behavior found in ants is that the information about a food source is mass communicated; that is, the number of workers that respond to a trail is determined in part by the number of workers laying a trail, which in turn is believed to be dictated by the quality of the food source. If this view is correct, then we might predict that when the bait is removed from the end of an active trail, there should be an "overshoot" where fresh workers continue to arrive at the bait area for several minutes.

Using the mealworm, cricket or honey bait set out in part A, attempt to test this hypothesis by keeping track of the buildup of workers at the bait following initial discovery. As soon as the total number of workers in a pencil-marked circle surrounding the bait reaches 10 - 12, gently remove the bait and note the time. Thereafter at 1 min intervals record the total number of ants inside the circle where the bait rested. Continue to count and record for

TABLE 16-1. Data sheet for study of chemical recruitment in ants.

PART A. Behavior of single foraging ants responding to 3 different
     foods.

Food Type	Behavioral responses of individual: ad libitum notes on food encounter behavior
Drosophila	
Honey	
Mealworm or cricket	

PART B. Trail disruption and directionality experiments.

Manipulation	Observations
1. One paper slid out of line	
2. One paper rotated 90°	
3. One paper reversed and realigned with trail	

PARTS C AND D. Bait removal and food quality experiments.

Bait type	Distance from bridge	No. of ants within bait circle at successive minute intervals following bait removal								
		start	1	2	3	4	5	6	.....etc.	30
Removal exps:										
mealworm or cricket										
honey										
Quality exps.										
pure honey										
dilute honey (1:5)										

15 min or so. If time permits repeat, placing the bait twice as far from the bridge. Using the meter stick, be sure to measure the distance from nest to bait and record this information. Later, graph these data, plotting number of ants on the vertical axis and time elapsed following bait removal on the horizontal axis.

D. Food Quality

Optimality theory predicts that individuals should forage preferentially at more "profitable" food sources (see exercises 9 and 10). Do fire ants respond differently to "good" and "poor" foods?
Prepare a dilute honey solution by mixing 1 part honey with 5 parts water. At equivalent distances from the bridge concurrently expose both this "poor" and and an equal amount of "good" (undiluted) honey to the ant colony (puts drops on cover slips). Observe both for 30 min, recording recruitment rates. Graphically compare your results, plotting the number of feeding individuals on the vertical axis and time in minutes from start of the experiment on the horizontal axis. If time permits note which food source is the first to be depleted.
Note: If desired, sucrose can be substituted for honey. Use solutions of 50% and 10% by volume. If time permits, student-generated experiments using other concentrations or foods often produce interesting results.
How might information on food quality be communicated? See the papers by Wilson (1962) and Hangartner (1969) for 2 suggestions.

SUGGESTIONS FOR FURTHER STUDY

1. Creation of Artificial Trails

Place a fresh food source some distance from the nest, in a spot unvisited by foragers. Grasp an ant with 2 pairs of fine-tipped forceps, 1 about the thorax and 1 about the abdominal petiole. Usually the sting will extrude when this is done. With a quick snap, sever the abdomen. Use the abdomen to streak a thin trail from the food source to an active trail. Do not allow your "applicator" to touch any workers. Observe the behavior of foraging ants encountering this artificial trail and compare it with that observed previously.

2. Ant Orientation to Trails

As an added challenge, place an anesthetized worker ant (see Aids to the Instructor for exercise 14 for anesthesia technique) under a dissecting microscope and cross its antennae. Carefully glue the antennae together using melted beeswax (Fig. 16-2). Using another ant, carefully coat one antennal club with beeswax. Taking a third ant, simply dab the base of each antenna with nail polish. (Why should this be done?)
When you have successfully prepared these ants (it does require some practice!), attempt to get the ants to follow an artificial trail. Compare their behaviors. What does this tell you about how a fire ant orients to its trail?

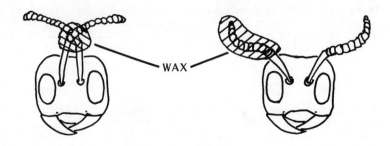

Fig. 16-2. Antennal manipulations showing glue placement used in suggestion 2 above. Apply beeswax melted with the tip of a heated dissecting needle to the places indicated.

SUMMARY QUESTIONS

1. What consistent differences in behavior are there between successful and unsuccessful foragers?

2. Predict a relationship between the time necessary to reach the maximum overshoot number and the distance of the bait from the nest.

3. What relationship might you expect between the number of ants involved in the maximum overshoot and each of the following variables? (a) bait quality; (b) bait quantity; (c) state of buildup when bait is removed; and (d) distance of bait from nest.

SELECTED REFERENCES

Barlin, M. R., M. S. Blum and J. M. Brand. 1976. Fire ant trail pheromones: analyses of species specificity after gas chromatographic fractionation. J. Insect Physiol. 22:839-844.
Hangartner, W. 1969. Structure and variability of the individual odor trail in Solenopsis (Formicidae). Zeit. Vgl. Physiol. 62: 110-120.
Taylor, F. W. 1978. Foraging behavior of ants: theoretical considerations. J. Theor. Biol. 71:541-565.
Wilson, E. O. 1962. Chemical communication among workers of the fire ant Solenopsis saevissima. 1. The organization of mass foraging. Anim. Behav. 10: 134-147.

17. Exploratory and Recruitment Trail Marking by Eastern Tent Caterpillars

T. D. Fitzgerald
S.U.N.Y., Cortland

Eastern tent caterpillars lay down exploratory trails as they move along the branches of the host tree in search of food. These trails enable the caterpillars to find their way back to the tent after feeding. In addition, successful foragers reinforce the trails that they follow back to the tent with a substance particularly stimulating to the caterpillars. Hungry tentmates follow these reinforced trails preferentially, and thus are recruited to food finds (Fitzgerald, 1976). The critical difference between exploratory and recruitment trails, however, has not yet been determined. Although the trails of the tent caterpillar are visible due to the accumulation of silk which is secreted by the larvae as they move along the branches, the caterpillars employ contact chemoreceptors rather than visual receptors when following their trails.

The following exercises demonstrate the chemical basis of trail following in this insect, and outline procedures for studying aspects of trail marking and recruitment under laboratory and field conditions. The activities fall into 2 categories: those involving studies of whole colonies in both the field and laboratory, and those involving laboratory studies of individual larvae. Studies of individual larvae require the least preparation and can be completed during a single laboratory period.

Although their natural field season is short, tent caterpillars can be reared easily following the procedures outlined in the Aids to the Instructor, and can be maintained under simulated field conditions in the laboratory.

METHODS

Subject

The eastern tent caterpillar, <u>Malacosoma</u> <u>americanum</u> (F.), occurs throughout eastern North America from Florida to New Brunswick (Stehr and Cook, 1968). Colonies consist of siblings which construct a communal silk tent in the branches of cherry or apple trees in the spring soon after eclosion. The caterpillars exhibit a recurring pattern of daily activity which is particularly pronounced under constant laboratory conditions (Fitzgerald, 1980). The onset of an activity period is marked by the movement of the caterpillars from the inner chambers of the tent to the surface, where they aggregate briefly while adding new silk to the structure. They then move, often en masse, to branches where they feed on the young leaves. The caterpillars return to the tent immediately after feeding and rest until the onset of the next activity period. Under field conditions, this pattern, though readily discernible, is less precise and is often completely disrupted by inclement weather, natural enemies, and over-exploitation of food. In the laboratory, colonies average approximately 4 activity periods per day, which occur at about 6 hr intervals.

The field season is short. Caterpillars chew through the chorion of the egg in early spring just as the buds break, and complete their larval development in 5 to 8 weeks. Mature larvae leave the tent and pupate in crevices, under rock overhangs, and in other protected locations. Adults emerge about 2 weeks later, and may mate and oviposit on the same day. Eggs are deposited in a tight cluster of 100 to 300 or more around a branch of the host tree, and are covered with a frothy layer of spulamine. Embryogenesis proceeds rapidly, and fully formed larvae appear in the eggs in 2 to 3 weeks. The eggs, however, must be exposed to a period of cold before the larvae eclose. Thus, in the field hatching does not occur until the following spring. Snodgrass (1924) provides a good discussion of the life history of this species.

The eastern tent caterpillar has a broad behavioral repertoire, and is an excellent subject for the study of insect behavior. While the exercises outlined below focus on the trail system established by the larvae as they move between their tent and feeding sites, other studies should suggest themselves to the alert student.

Species other than M. americanum may also be used for the laboratory part of the exercise. The forest tent caterpillar, M. disstria Hbn., and the ugly nest caterpillar, Archips cerasivoranus (Fitch) follow trails (Fitzgerald and Edgerly, 1979); the smaller instars are good laboratory subjects. The gypsy moth, Porthetria dispar (L.), and the fall webworm, Hypantria cunea (Drury), however, make poor laboratory subjects for these exercises. A study of the European birch tent caterpillar, Eriogaster lanestris L., by Weyh and Maschwitz (1978) suggests that many of the activities suggested here are also applicable to this and other species of tent caterpillars.

Materials Needed

Cherry or apple leaves; rearing apparatus (Fig. 17-1); filter paper; razor blades; hexane, ether, or similar solvent; high intensity lamp (or microscope illuminator); index cards; disposable 5 μl micropipets or capillary tubes, or a gas chromatograph syringe which dispenses 5 μl; Eberhard Faber Microtomic 4B pencil and other pencils; sandpaper.

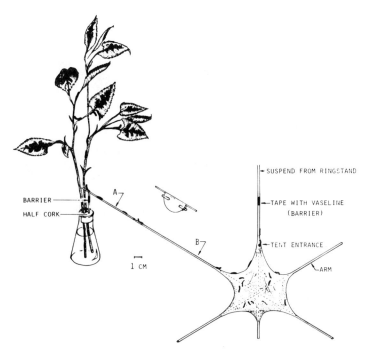

Fig. 17-1. An apparatus used to rear whole colonies of tent caterpillars under simulated field conditions in the laboratory. The structure is constructed from 3/16 inch wooden dowels fastened together with hot glue. Along with their accumulated silk, colonies of first or second instar larvae are placed at the center of the stand, where they will construct a tent. From this time until they are fully grown, caterpillars will move out to the branches to feed and return to the tent after each foraging session. See text for additional details.

Procedure

A. Field Studies of Recruitment

This study can be easily conducted in the field during the spring. Locate a small tree containing a colony of tent caterpillars and defoliate it. Over the next 24 hrs., make periodic observations of the intensive exploratory behavior exhibited by the caterpillars as they comb the branches of the host tree in search of food. When most of the branches have been covered with silk trails, locate an individual caterpillar that is searching at a considerable distance from the main aggregation. Pin a fresh leaf in its path so that it discovers the food. After feeding, the larvae will return to the tent. Note the precise path taken by the caterpillar as it moves over the branches of the host, then observe the responses of the hungry caterpillars to this pathway.

Since it is not presently clear whether larvae will invariably lay down a recruitment trail under these circumstances, it may be necessary to repeat the procedure if the colony does not respond. Observations can be quantified by comparing the number of caterpillars that spontaneously discover the feeding site before the first caterpillar returns and the number that discover it during an equivalent period of time after the first caterpillar returns. Alternatively, several potential experimental feeding sites can be selected at the outset of the study. One of these can be randomly designated the actual feeding site. The number of larvae that visit this site after the return of the fed caterpillar can be compared with the number that visit the other sites during an equivalent period.

Late in the field season as the caterpillars near maturity, it is common to find trees already defoliated and full of hungry, searching caterpillars. Such trees lend themselves to immediate experimentation. See the sections below for other studies that can be adapted to the field.

B. Laboratory Studies of Trail Marking and Recruitment

If your instructor has not already done so, establish a laboratory colony of tent caterpillars following the guidelines in Aids to the Instructor. Transfer the caterpillars to a rearing stand (Fig. 17-1) to simulate field conditions. The caterpillars show strong attachment to their silk, and will construct a tent at the transfer site. (Colonies of small larvae collected from the field can also be transferred in the same manner.) The caterpillars usually discover the leaves with little difficulty, but the process can be speeded up by temporarily placing a light at tent level behind the leaves. (Occasionally the larvae will attempt to establish a new tent on the host branches soon after transfer. If this happens, gather up the larvae with their accumulated silk, and return them to the center of the stand.) Once the colony has formed its tent at the transfer point, the caterpillars will return to it after each forage until they are fully grown. Provide the caterpillars with a constant supply of branches bearing fresh, young leaves.

It will become apparent that the caterpillars are reluctant to move over areas not previously covered with their silk trails. You can investigate the way in which caterpillars establish new trails by covering a section of their trail with a sheet of filter paper as shown in Fig. 17-1. If the sheet is put in place while some of the caterpillars are out feeding, you can observe the response of both fed, returning caterpillars and hungry, outbound larvae to the trail-less substrate. Are there apparent "leaders" that push across these gaps, or do the larvae push across in groups? How many larvae must cross before the caterpillars move across without hesitation?

When a trail has been established on the filter paper, remove the paper by carefully cutting the strands of silk with a razor blade at both ends where they join the bridge. Rinse the filter paper <u>thoroughly</u> in 2 changes of hexane, ether, or similar solvent. Allow the solvent to evaporate, then reattach the sheet to the bridge. How do the caterpillars react to the extracted trail? Are the tactile or visual properties of the silk adequate to elicit the following response? See Section C below for techniques you can use to replace the pheromone.

Recruitment can be studied in the laboratory, but requires careful preparation and timing. Working with a colony that has been established on a stand, detach section A of the bridge so that the caterpillars cannot reach the leaves. The hungry caterpillars will then establish new exploratory trails on the arms of the stand as they search for food. When it is apparent that they are moving over the arms with ease, join the tip of bridge section A to section B, and allow 4 or 5 caterpillars to cross over. Immediately separate the bridge sections and allow the isolated group of caterpillars to feed. After feeding, the caterpillars will return to the bridge (10-20 min.). When the first caterpillar reaches the bridge, abut the end of the bridge to 1 of the 3 horizontal arms of the stand selected randomly, and allow the fed caterpillars to return. Immediately place branches at the ends of the other 2 arms. Record the number of caterpillars that move to each of the 3 arms. Since recruitment trails are formed on the tent surface as well as on the branches of the host tree, you should carefully observe the paths taken by the first of the fed returning larvae, and note how successive caterpillars, both fed and hungry, respond to these tent trails.

Fed, returning caterpillars are very sensitive to disturbances and abrupt changes in trail strength or quality. If the arm selected to return the caterpillars is not strongly marked with fresh exploratory trails, the caterpillars initially may refuse to cross onto it. Also, there may be too many caterpillars active at 1 time to carry out this recruitment test without ambiguity. To reduce confusion, the colony can often be quieted by directing the light from a high intensity lamp or microscope illuminator onto the tent surface. The caterpillars react to such light by aggregating in a motionless clump under it. If this is done soon after the initial contingent of larvae are collected, the colony should be considerably less active when the fed larvae are returned. The light can then be removed, and the responses tallied.

C. Studies with Individual Larvae

Many aspects of trail following behavior can be investigated with individual larvae and simulated trails laid down on index cards. Prepare a trail extract by soaking approximately 5 mg of fresh tent or trail silk per 0.1 ml of hexane, ether, or other solvent for 5 min. Agitate occasionally to enhance extraction. Five µl of the extract laid out in a narrow line 20 mm long on an index card will elicit trail following. (Filter paper or other highly absorbent material should not be used.) Do not draw a pencil line on the card; rather, indicate the ends of the extract trail with dots.

Lay the extract down slowly to minimize lateral diffusion of the solvent. Only first to third instar larvae (larvae less than 7 mm in length) should be used for these tests. Older larvae follow trails just as readily as younger larvae under natural conditions, but are easily disturbed and often wander when placed on cards. The following suggested studies represent only a few of those that might be conducted, and emphasize basic features of trail following. Groups of students should share their results in order to generate a sufficient data base.

1. Demonstration of Chemical Trail-Following Behavior

Lay down an extract trail as indicated above, and a control trail consisting of the solvent alone. Place individual larvae at 1 end of each trail, and carefully observe their behavior. Illustrate the path taken by each of the larvae relative to the trail. Repeat this process several times, and compare your results with those of your classmates.

2. Stimulus Specificity

It has been suggested that in the absence of an authentic trail substance insects may follow trails of chemicals that they would not follow ordinarily. Such lack of specificity, if it should occur, could lead to erroneous conclusions regarding the nature of the trail-following stimulus. To determine how specifically the caterpillars respond to their trail substance under the conditions of this bioassay, wash host leaves and bark in separate containers of solvent and lay down trails of these. (Your instructor may provide these.) Compare the responses of caterpillars to these trails and to trails prepared from silk extracts. You can quantify your observations by recording the time it takes each larva to reach the distant end of the trail. Repeat several times, and share your results with other groups. Record all your observations.

3. Trail Persistence and Larval Discrimination of Trail Age

Artificial trails can be arranged in the form of a "Y" (Fig. 17-2) to determine whether the caterpillars can discriminate between old and new trails. Prepare an extract of fresh tent or trail silk as indicated above. Eight to 12 hrs. before the exercise is to be conducted, lay down 5 µl of the extract along 1 of the arms of the "Y". Prepare a total of 10 "Y"s in this manner, alternating the side to which the extract is applied. Store the cards in the open at room temperature. (Your instructor may supply you with cards on which 1 arm has already been treated.)

Just prior to conducting the test, apply 5 µl of the same extract to the untreated arm, and a like quantity to the runway. Place a caterpillar at the end of the runway and record the arm it takes when it reaches the choice point. Use care when laying down the trails. Make them as narrow as possible to minimize the amount of overlap at the junction of the arms and the runway. If the caterpillars follow only the most recently deposited trail,

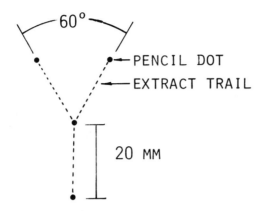

Fig. 17-2. Procedure for laying out trail extract for 2-choice bio-assay. See text for additional details.

determine whether the aged trail is still attractive by placing a larva at the outer end of the arm bearing the aged extract and observe its response to the trail.

Record your results and share your data with classmates.

4. Trail Reinforcement

Trail-making ants and termites may or may not reinforce trails as they move over them. It is important to determine whether or not this is the case, since it can influence the interpretation of investigations of trail-marking behavior. The following test can be conducted to determine whether caterpillars reinforce their trails under the conditions of this bioassay.

Outline a "Y" with dots as shown in Fig. 17-2. Lay down 5 µl of extract along the runway and another 5 µl on 1 arm. Place a caterpillar at the end of the runway and let it move up and down the trail 10 times. Remove the larva. (Do not pick the larva up while it is still on the trail. Wait until it reaches 1 end of the trail, and push it off the trail by prodding it on the tip of the abdomen with a blunt object before lifting it.)

To determine whether the caterpillars are reinforcing the trail, lay down 5 µl of the original extract on the untreated arm of the "Y". Place a larva at the end of the runway and note the arm it selects. Repeat 5 times, and share your results with other groups. Record your results.

5. The "Pencil Factor"

Pheromones may be unique chemical substances produced exclusively by a given insect or a relatively common compound adapted to a communicative function by an insect. Pheromones of the latter type may turn up in unexpected places. Certain pencil leads contain a chemical factor which elicits trail following in the eastern tent caterpillar. Whether this substance is the authentic trail substance or merely similar to it (a parapheromone) has not yet been determined. The active chemical is a component of additives of plant or animal origin that are used in lead formations. Bioassays of a number of different brands of pencil leads have shown that some are wholly inactive, whereas others (such as Eberhard Faber Microtomic 4B) elicit trail-following behavior.

Using the pencil provided by the instructor, demonstrate the occurrence of a chemical trail factor by drawing a 20 mm line on a card and observing a larva's response to it. Extract the trail by rinsing the card in several changes of hexane or similar solvent. Retest the line. Are the visual components of the trail adequate to elicit trail following?

If time is available, prepare extracts of lead and observe the caterpillars' reponse to the extract trails. (Lead can be abraded on sandpaper before extracting.) You can also bioassay different brands of pencils and different degrees of hardness of the same brand of your own and classmates' pencils. If you find several that elicit the response, arrange the lines in the form of a "Y" and perform tests similar to those outlined in the section above to determine which produces the strongest response. Record your results.

SUMMARY QUESTIONS

1. Why do tent caterpillars employ both exploratory and recruitment trails?

2. The trails of the tent caterpillar are more persistent than those of many trail-following ants and termites. Speculate on the reasons why this is the case.

3. Does recruitment appear to serve the same purpose in tent caterpillars as it does in ants and termites? Can the behavior in tent caterpillars properly be termed recruitment?

4. Individual colonies of tent caterpillars are aggregations of siblings. What bearing might this have on the evolution of recruitment in this species?

5. Speculate on the ways in which tent caterpillars might differentiate exploratory and recruitment trails.

SELECTED REFERENCES

Bucher, G.E. 1959. Winter rearing of tent caterpillars, Malacosoma spp. (Lepidoptera: Lasiocampidae). Can. Entomol. 100: 411-416.
Fitzgerald, T.D. 1976. Trail marking by the larvae of the eastern tent caterpillar. Science 154: 971-973.
Fitzgerald, T.D. and J.S. Edgerly. 1979. Specificity of trail markers of forest and eastern tent caterpillars. J. Chem. Ecol. 5: 565-574.

Fitzgerald, T. D. 1980. An analysis of daily foraging patterns of laboratory colonies of the eastern tent caterpillar, <u>Malacosoma americanum</u> (Lepidoptera: Lasiocampidae), recorded photoelectronically. Can. Entomol. 112: 731-738.

Snodgrass, R.E. 1924. The tent caterpillar. Annual Rept. Smithsonian Inst. 2724: 329-362.

Stehr, F.W. and E.F. Cook. 1968. A revision of the genus <u>Malacosoma</u> Hubner in North America (Lepidoptera: Lasiocampidae): Systematics, biology, immatures, and parasites. Bull. 276, Smithsonian Institution.

Weyh, R. and U. Maschwitz. 1978. Trail substance in larvae of <u>Eriogaster</u> <u>lanestris</u> L. Naturwisenschaften 65: 64.

18. Dominance in a Cockroach (*Nauphoeta*)

George C. Eickwort
Cornell University

Dominance -- the ranking of individuals on the basis of real or apparent authority, strength, influence, etc. -- is less well documented for invertebrates than it is for vertebrates. However, a number of insects as diverse as social wasps, crickets, dragonflies, and burying beetles are known to form dominant-subordinate relationships that may result in a linear, fairly stable hierarchy, at least under experimental conditions. Individual recognition generally does not appear to play an important role for invertebrates. Rather, being defeated by a more dominant individual seems to bring about some internal (possibly hormonal) change that makes the appearance of subsequent subordinate behavior more likely.

Males insects that form dominance hierarchies are usually territorial. Usually, only highest ranking males (alphas) retain the "better" territories for long periods of time. Lower ranking males (betas) occupy lower quality territories or irregularly hold "better" territories, from which they can be chased with relative ease by patrolling alphas. Lowest ranking males (gammas) usually have no territorial rights. The ownership of a "good" territory increases the mating success of a male.

In some insects, as in most vertebrates, when animals are close in rank an animal is more likely to win on its own territory. However, in Nauphoeta roaches the outcome of a fight depends solely on rank. Males roaches gaining in rank tend to acquire territories or become better established in them; those losing rank tend also to lose their territories. The hierarchy established is semilinear; a dominant male does not always defeat those subordinate to him, but he does win over 50% of the encounters.

The purpose of this exercise is to determine the role of aggression in the development and maintenance of the hierarchy that characterizes the social organization of a group of adult male cockroaches. We will examine the different behaviors exhibited by dominant and subordinate roaches when they interact, the formation of a hierarchy, and the responses of dominant and subordinate males to the introduction of new males and virgin females.

120

METHODS

Subject

The cockroach Nauphoeta cinerea (Blattodea: Blaberidae) is a native of Africa that has become established in human domiciles throughout tropical and subtropical regions, including Florida. It is easily maintained in culture tanks on standard laboratory animal diets. Like most roaches, it is nocturnal. Sexes are easily distinguished (Fig. 18-1). Females are ovoviviparous. Newly hatched nymphs remain with their mother for a day or less, and crawl under her wings where they are protected while their cuticle becomes sclerotized. Nymphs and adult females do not form dominance hierarchies.

If Nauphoeta are unavailable, with modification it may be possible to conduct this exercise with other roach species. Males in species of 3 other blaberid genera are territorial -- Blaberus, Gromphadorhina, and Eublaberus (Gorton et al., 1979, and references therein). Male aggression and at least transient territoriality have also been observed in Periplaneta (Bell and Sams, 1973). Blattella appears to be less social and to lack definite hierarchies (Breed et al., 1975), but is useful for the courtship portion of the exercise (Bell et al., 1978).

Materials Needed

Colony of Nauphoeta maintained in culture tanks (such as plexiglass boxes); white quick-drying enamel (model airplane dope) or acrylic paint; fine brushes; petri dishes; wax pencils; carbon dioxide; stereomicroscope; red light bulbs.

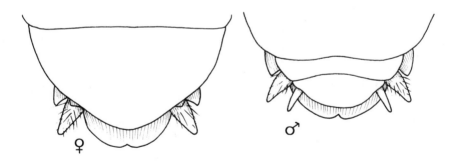

Fig. 18-1. Ventral view of last nymphal instar male and female Nauphoeta roaches. males have a short subgenital plate (apical abdominal sternite) with a pair of short styli in addition to the longer cerci. Females have a large subgenital plate and possess cerci but lack styli.

Procedure

A. Behavioral Interactions Within an Established Dominance Hierarchy

For each pair of students, a petri dish with 3 differently marked male roaches is available. These roaches were kept in solitary isolation until 3 days prior to the exercise, when they were introduced into the petri dish. At that time they began to establish a dominance hierarchy, which would become stable in about 9 days. By day 3, when this exercise is conducted, the alpha male can usually be identified and male-male encounters are at their peak of frequency. Later, male-male encounters will become less violent and less frequent as the hierarchy becomes more stable. However, shifts in the dominance order will still occur, accompanied by brief periods of aggression between the individuals involved in the reshuffling.

With a partner, obtain a petri dish of roaches, and observe the interactions as the roaches settle down. How do they groom themselves? When 2 roaches meet, how do they respond to each other? What sense organs are involved? Aggressive behaviors may include (1) approach, a decisive and direct movement of 1 individual toward another; (2) lunge, a quick forward movement of 1 individual toward another, terminating in contact; (3) stilt posture, an individual elevates his body off the dish, especially with his forelegs, sometimes accompanied by jerking or abdominal extension; (4) climb, 1 individual climbs onto the abdomen of another; and (5) butt, 1 individual executes a quick thrust with his pronotum against the body of another. Rarely, aggressive males will grapple or bite other males. Submissive behaviors include (6) crouch, lowering the body against the dish with legs tucked beneath wing margins; and (7) retreat, withdrawal after contact (terminology from Bell and Gorton, 1978). Which classes of behavior are most common?

After a period of informal observation, begin to assess the relative frequency of aggressive vs. submissive behaviors exhibited by each member of each possible pair as they meet. As 1 partner observes, the other should record encounter outcomes on the data sheet (Table 18-1). Midway through, exchange roles. If encounters between 2 given males are not followed by a fight, assign a tentative rank based on which male shows avoidance of the other, or through their encounters with other animals of known rank. If the males do not move, shake the dish. Observe for at least 15 minutes, or until you have sufficient encounters in each category to construct a tentative dominance hierarchy for your 3 male roaches.

The dishes are too small to allow the formation of discrete territories, but some aspects of territorial behavior should be apparent. Estimate which of the males is most frequently in motion. This patrolling behavior is characteristic of a male that "owns" a territory. Where does he fit in the dominance hierarchy?

B. Behavioral Interactions Between Dominant Roaches

Students now work in groups of 4. One pair of students contributes the dominant roach from their dish to the other dish.

TABLE 18-1. Determining dominance/subordinance in the cockroach, Nauphoeta. Data sheet for culture with established dominance hierarchy; repeat for culture with introduced fourth roach (alpha male), and for culture of newly introduced males.

Roach No.	Mark	Laboratory Partners:
1		
2		
3		
4	(newcomer introduced from other culture)	

RAW DATA: Keep a running tally using shorthand notation like this: 1 > 3, 2 > 3, 2 > 1, etc. where 1 > 3 means an encounter between roaches 1 and 3 was won by roach 1.

TALLY:

Total No. of encounters:

DATA REDUCTION: First enter total wins and losses for all possible combinations of individuals in a matrix organized like this (winners in rows, losers in columns);

Example:		loser			Your data:		loser			
		1	2	3			1	2	3	4
	1.	-	2	7		1.				
winner	2.	1	-	0	winner	2.				
	3.	0	2	-		3.				
						4.				

Then if necessary rearrange into a new matrix which minimizes the scores below the diagonal (i.e., try to get most of the zero scores below the diagonal). Scores that are greater than zero and remain below the diagonal after this rearrangement represent cases of reversal. Note: there are several possible arrangements of 3 or 4 individuals in a matrix; the 1 with the fewest reversals below the diagonal is assumed to represent the hierarchy.

Rearranged		loser		
Example:		1	3	2
	1.	-	7	2
winner	3.	0	-	2
	2.	1	0	-

TENTATIVE LINEAR HIERARCHY: For original 3 roaches:
(Example: 1 > 3 > 2)

COMMENTS:

(Note differences in body size or shape of paint mark if 2 males possess the same mark.) Observe the interactions of the introduced male with each of the other resident males, and record them on the data sheet (Table 18-1). Note especially the interactions between the 2 dominant males. Which male is now dominant? Are interactions more or less frequent, and more or less violent, than those that that previously characterized the roaches in the established hierarchy?

At some time during these experiments, a male may raise his forewings, exposing his abdominal terga. This is <u>homosexual</u> behavior. A second male may crawl onto the exposed abdomen, exhibiting <u>pseudofemale</u> behavior. These sexual behaviors are more common when strange males are newly brought together. Their significance is unknown.

At the end of this experiment, return the introduced male to his own dish, and allow the roaches to settle down.

C. Behavioral Interactions Among Newly Introduced Males

Each pair of students obtains a petri dish in which 3 differently marked males, previously kept isolated, have just been placed. Observe their interactions as the roaches settle down. On a second data sheet, record the relative frequency of aggressive vs. submissive behaviors exhibited by each member of a pair when they meet, as you did in part A. Are interactions more or less frequent and/or violent than those observed in part A? Is the hierarchy as well established? How can you measure this? How many interactions occur before you are sure that you can identify the alpha roach?

Transfer the alpha roach from the established hierarchy into this dish, and record interactions as in part B. Return this roach to his own dish, then transfer the alpha roach from the new dish to the established hierarchy. Record interactions, and compare results with those of the reverse transfer and of part B.

When Ewing (1967) mounted a male roach on a stick and used it to "attack" newly emerged males, she found that some males gradually became sluggish, stiff and semi-paralyzed. Eventually these males died from this "stress syndrome". Similar deaths were observed in unmanipulated cultures of newly emerged males, although older males were more immune to the stress syndrome. Other experiments indicate that some physiological changes occurred in animals when they lost fights. Even among those subordinate males that did not eventually die, the stressed animals showed a much lower increase in weight than did dominant roaches, and some actually lost weight. These physiological changes may have been induced by neurohormones released by the corpora cardiaca.

D. The Influence of Females on Male Behavior

Introduce a virgin female into the petri dish containing the established hierarchy. (She will be recognizable by the lack of paint mark.) Observe the interactions of each male with her. Does either sex exhibit aggressive behavior like that between males? Does the alpha male attempt to establish dominance over the female?

Does the frequency of male-male interactions increase after the female is introduced? Is the dominance order maintained?

Soon, 1 or more males should begin courtship by raising his wings and exposing his abdominal terga, thus releasing a pheromone called "seducin". If the female is unreceptive, the male may lower his wings and stridulate, producing a faint sound which is audible only if the dish is uncovered. The sound is produced by rubbing the ventral posterior edge of the pronotum against the upper surface of a forewing, coupled with rhythmic pumping and extension of the abdomen that brings the forewing into the correct position.

If the female is receptive, she will climb onto the male's abdomen and "feed" on his terga, and copulation will ensue. Unfortunately for purposes of this study, not all virgin females are receptive. Very young (less than 4 days post-emergence) and very old females will not mate. Also, virgin females often require a period of acclimatization in the dishes before they will mate. What is the reaction of the other males to a copulating pair? Which male is successful in copulating? Pool the class results to determine whether there is a significant correlation between dominance rank and mating success. Can you hypothesize an explanation for the results? (Remember that your experiment is carried out under unnaturally crowded conditions where subordinate males cannot be excluded from territories.)

Fukui and Takahashi (1980) have shown that males can discriminate females from males on the basis of a surface pheromone that they perceive with their antennae. Breed et al. (1980) found that females can discriminate dominant from subordinate males on the basis of odor.

SUGGESTIONS FOR FURTHER STUDY

The excellent studies of Leonie S. Ewing of the University of Edinburgh have formed the basis for this exercise, and ideas for further study can be found in her papers cited in the references. To test the effects of artificial stress on a dominance hierarchy, you may wish to individually chill roaches after they have established a dominance hierarchy, then reunite them after they recover (Ewing and Ewing, 1973). You could test the stability of dominance hierarchies by separating pairs of roaches for varied lengths of time and then reuniting them. What happens if each roach is then paired with an animal of different rank than his original partner? Design experiments to test the contribution of size, age, or nutritional status in determining the status of roaches in a hierarchy. Ewing (1973) describes a procedure by which roaches will establish semi-natural territories in a large cage, so the interaction of territoriality, dominance status, and mating success can be studied with different densities of males and females.

SUMMARY QUESTIONS

1. In the laboratory you have constructed a linear hierarchy for male Nauphoeta. Would you expect the hierarchy also to be linear under natural conditions? Why or why not?

2. What appears to be the main selective advantage of a male attaining dominant status in Nauphoeta? How might your answer differ if you observed roaches under natural conditions?

3. What advantages are there in using 3 marked individuals in a dish to study dominance interactions?

SELECTED REFERENCES

Bell, W.J. and R.E. Gorton, Jr. 1978. Informational analysis of agonistic behavior and dominance hierarchy formation in a cockroach, Nauphoeta cinerea. Behaviour 67: 217-235.

Bell, W.J. and G.R. Sams. 1973. Aggressiveness in the cockroach Periplaneta americana (Orthoptera: Blattidae). Behav. Biol. 9: 581-593.

Bell, W.J., S.B. Vuturo and M. Bennett. 1978. Endokinetic turning and programmed courtship acts of the male German cockroach. J. Insect Physiol. 24: 369-374.

Breed, M.D., C.M. Hinkle, and W.J. Bell. 1975. Agonistic behavior in the German Cockroach, Blatella germanica. Z. Tierpsych. 39: 24-32.

Breed, M. D., S. K. Smith, and B. G. Gall. 1980. Systems of mate selection in a cockroach species with male dominance hierarchies. Anim. Behav. 28: 130-134.

Ewing, L.S. 1967. Fighting and death from stress in a cockroach. Science 155: 1035-1036.

Ewing, L.S. 1972. Hierarchy and its relation to territory in the cockroach Nauphoeta cinerea. Behaviour 42: 152-174.

Ewing, L.S. 1973. Territoriality and the influence of females on the spacing of males in the cockroach, Nauphoeta cinerea. Behaviour 45: 282-303.

Ewing, L.S. and A.W. Ewing. 1973. Correlates of subordinate behaviour in the cockroach, Nauphoeta cinerea. Anim. Behav. 21: 571-578.

Fukui, M. and S. Takahashi. 1980. Studies on the mating behavior of the cockroach, Nauphoeta cinerea Olivier 1. Sex discrimination by males. Appl. Entomol. Zool. 15: 20-26.

Gorton, R.E., Jr., J. Fulmer, and W.J. Bell. 1979. Spacing patterns and dominance in the cockroach, Eublaberus posticus Dic (Dictyoptera: Blaberidae). J. Kans. Entomol. Soc. 52: 334-343.

Hartman, H.B. and L.M. Roth. 1967. Stridulation by the cockroach Nauphoeta cinerea during courtship behaviour. J. Insect Physiol. 13: 579-586.

19. Cricket Phonotaxis: Experimental Analysis of an Acoustic Response

Glenn K. Morris and Paul D. Bell
Erindale College,
University of Toronto

Phonotaxis is the oriented locomotion of an animal in response to a source of airborne sound; it may be positive (toward the stimulus) or negative (away). Commonly phonotaxis takes the form of a direct locomotory response by a silent female to a single stationary male's calling song.

In 1913, Johann Regen carried out the first demonstration of phonotaxis in an animal. He created a model of a male cricket which consisted only of its normal calling song, and showed that sexually receptive females would approach this model. A "telephone", placed in an arena, was connected to a microphone in another room; this arrangement could be used to relay the stridulations of a remote caged male. Within the arena, 30 cm from the telephone, Regen placed a second male in a small cage. A female was then introduced 1 m away at the other end of the arena. She moved to the caged, singing arena male (revealing her responsiveness). He was then silenced, and the telephone "song-relay" turned on. The female responded by moving to within 1 cm of the phone. In this fashion, Regen established that the locomotion of a cricket female toward a male's song can be elicited by a purely acoustic stimulus.

Male crickets (Orthoptera: Gryllidae) produce sound by stridulation -- rubbing 1 body part against another (Fig. 19-1). Females are silent. The acoustic communication system is actually quite complex, involving several different songs. However, the loudest and most commonly heard is the "calling song".

The calling song of the house cricket, Acheta domesticus (L.), is a series of chirps repeated at fairly regular intervals. Each chirp is composed of 2 and sometimes 3 phonatomes. A phonatome is all the sound produced during 1 cycle of movement of the generating mechanism (the forewings, or tegmina, in the case of crickets). Fig. 19-2 shows the sound structure of a cricket song.

In this exercise, we will be studying phonotactic behavior in the house cricket by measuring female cricket response on exposure to 3 experimental treatments: a broadcast recording of the normal calling song of the male, a recorded synthetic sound produced by a wave generator, and silence, the absence of acoustic stimuli to serve as a control. Following appropriate randomization, trial procedure, and statistical analysis, results will be interpreted and written up as though for submission to a scientific journal.

127

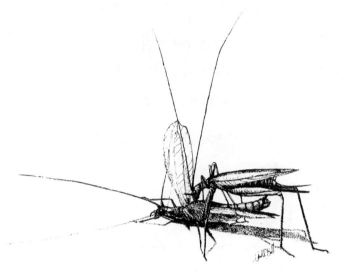

Fig. 19-1. A tree cricket singing during courtship; the female feeds on secretions of the male's metanotal gland. Note the elevated position of the front wings. Sound is produced when the sharp edge (scraper) at the base of one front wing is rubbed along a ridge (file) on the ventral side of the other front wing. At rest, the right wing is usually uppermost; while both wings possess scraper and file, the file is usually longer in the upper wing and the scraper better developed on the lower.

METHODS

Subject

The house cricket, _Acheta domesticus_ is widely available from biological supply companies and easily maintained in culture. Species of _Gryllus_, the common black field cricket, or of _Oecanthus_ (Fig. 19-1), tree crickets, could also be used. Prior to this exercise, the crickets will have been separated by sex, and females (recognizable by the cylindrical ovipositor) housed individually in small jars.

Materials Needed

Two audio-range tape recorders, each with a previously prepared treatment tape; 1 speaker switching unit (see Fig. 19-3); 15-20 plastic baby food jars or other small containers, each with an adult female cricket; container with labelled pennies used for randomization of treatment and specimen number; thermometer; stopwatch. Optional: sound level meter.

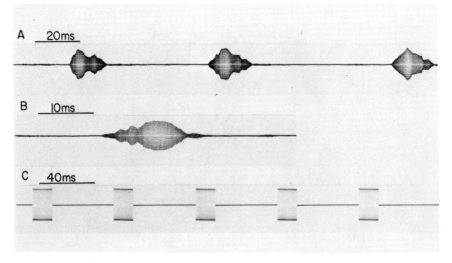

Fig. 19-2. The song of the house cricket, <u>Acheta</u> <u>domesticus</u>. the chirp in this species consists of two or three phonatomes, most of the sound in each phonatome produced as the wings move together. (A) An oscillogram of one 3-phonatome chirp. (B) A single phonatome at a higher sweep rate of the oscilloscope beam showing the sinewave structure of the emitted sound. (C) An artificial signal from a wave generator: bursts of 4500 cycle sinewaves, close to the phonatome rate within the cricket's chirp, are repeated indefinitely.

Procedure

For this exercise, work in pairs or individually. In order to minimize duplication of expensive equipment, you will probably be required to work in shifts outside of the normal laboratory period. Arrange a time with the instructor or technician for access to the equipment; you will probably need 2 to 3 hours.

A. Randomization

Each cricket-containing jar is numbered; consider this number to be that assigned to the specimen in the jar. Obtain the box full of pennies; check to make sure it contains 1 penny for each treatment-specimen number combination (Treatments I, II, III and specimen numbers 1, 2, 3. . .15-20). Mix the pennies and then select 1 at a time (without replacement). Write each treatment-specimen number on a sheet of paper in the sequence drawn. The same individual may not be in 2 successive trials; if you draw the same specimen number twice in succession, return the second draw to the container, mix the pennies, and draw again. Continue until all pennies have been drawn and tabulated.

Consult the random number table (Appendix 2), and assign speaker positions to each of the treatment-specimen number combinations. Each speaker position is numbered 1, 2, 3, or 4. Once you have gained access to some point in the table (by stabbing

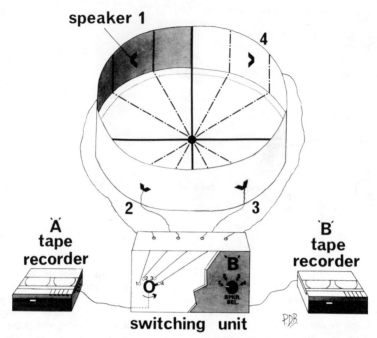

speaker 1

`A´ tape recorder

`B´ tape recorder

switching unit

Fig. 19-3. Equipment set-up for studying cricket phonotaxis. Each trial begins with a female cricket placed in center of the arena, and terminates as she crosses the exit line. The arena is 140 cm in diameter including a 10 cm exit border (dotted line). The wall's inner face is lined with sound absorbent material (e.g. fiberglass insulation) to reduce sound reflection. A switching unit permits playback through any of the 4 speakers for recorders "A" and/or "B".

with your finger, etc.), simply work your way along a row of digits, assigning 1, 2, 3, 4 as they occur and ignoring 5-9 and 0. Eventually you will emerge with a randomized sheet such as shown below:

Trial	Specimen Number	Treatment	Speaker
1	14	II	4
2	8	II	2
3	3	I	2

. . . . .etc.

B. Trial Procedure

Obtain the equipment from the instructor or technician, and familiarize yourself with its operation (see Fig. 19-3). Be careful not to move the taped volume knob on the tape recorders.
Prepare 3 tally sheets; for each, draw a large circle on a piece of paper and mark it off into 12 equal sectors. Indicate speaker locations.

During the trials, minimize outside disturbance to the crickets. Station yourself and your partner at diagonally opposite locations alongside the arena, at locations which are constant from trial to trial and which afford between you an uninterrupted view of the arena floor. This avoids any necessity to lean over the arena wall to follow the cricket's movements. Avoid movement, talking, bumping arena supports, etc. while the trial is in progress.

1. Consult the randomization sheet, rewind the tape of the appropriate recorder if necessary, and set the switching unit for playback from the indicated speaker. Check and record room temperature; crickets vary their phonatome rate with temperature, and the temperature of the tested female must be close to that of the recorded male.

2. Select the appropriately numbered specimen, loosen the jar cap, and invert the jar in the palm of your hand so that the cricket is resting on the inside of the lid and the jar is ready to be lifted away from the lid in a single movement.

3. Place jar and lid (lid downward) at the center of the arena and lift away the jar. As soon as more than half of the cricket's body is over the edge of the lid, depress the start control of the recorder and initiate speaker playback. Simultaneously begin timing with the stopwatch.

4. Continue the trial until the cricket first crosses the 60 cm line. At this point, stop playback and timer. Map the sector in which the 60 cm crossing occurs, estimating the exit point by eye as precisely as you can and marking it on the tally sheet.

5. To recapture the cricket, retrieve the cap from the arena center. With 1 hand hold the open end of the jar in front of the cricket; with the other hand, herd the insect until it runs or hops into the mouth of the jar. Turn the jar over, and when the cricket is in the bottom, screw the cap on. <u>Do</u> <u>not</u> <u>clap</u> <u>the</u> <u>jar</u> <u>over</u> <u>the</u> <u>cricket</u> <u>from</u> <u>above</u>! This frequently severs limbs, and legless crickets do not exhibit phonotaxis!

6. Check off the completed trial, and go on with the next.

C. Discussion

Treatment I is the normal calling song of the male house cricket (Fig. 19-2A,B). Treatment II is an entirely synthetic sound produced by a signal generator (Fig. 19-2C). A pure tone frequency is modulated or "gated" into a succession of sound pulses. This "manufactured" cricket call matches the normal song in carrier frequency (4500 cycles/sec.; 4.5 kHz) and in pulse duration (15 millisec) but is in other respects quite different. The most obvious difference is the lack of any chirp structure. The pulses simply repeat indefinitely at a rate of 180/sec. Treatment III is silence -- the absence of acoustic stimuli -- and may be considered a control.

The instructor will have pre-set the intensity level of the playback so that the cricket hears the sound model (at the arena center) with a consistent loudness between experiments, replicates, or treatments. A workable sound level would be about 55 dB (re 20 μPa, A weighting network) measured with a standard sound level meter. Other factors being equal, females respond more strongly to more powerful stimuli. Treatment II involves more sound energy per unit time than the normal chirping song. Therefore, equal intensity broadcast level for the 2 treatments requires that each artificial pulse have a lower amplitude. Stimulus incidence rate ("duty cycle") can be even more important with 2-speaker choice designs.

For each treatment, we advance the same null hypothesis: no effect of the acoustic model on the direction taken by the cricket in exiting from the arena. If the sounds we broadcast are attractive, we may expect a significantly higher number of exits within the speaker quadrant. But if our null hypothesis holds, then we would expect a roughly equal number of exits from each of the 4 quadrants. Thus if we utilized 19 crickets in our series of trials, for example, then for any 1 treatment we would expect 19/4 = 4.74 exits in each quadrant. This hypothesis can be tested by applying a chi square goodness-of-fit-test (see Appendix 1.2) to each treatment in turn.

You may also test the hypothesis that locomotion occurs more rapidly in the presence of sound than in silence. That is, are "times to first exit" significantly different between treatments I and III? Use either Student's t test or the Mann-Whitney U test (Appendix 1.4).

Write up and interpret your results as though you were preparing a paper for submission to a scientific journal. Indicate which journal you have selected, and conform to its editorial format. Though the data have been gathered jointly, each student must analyze and interpret them separately and submit a separate "paper". Consult your instructor or laboratory assistant if you have any problems in applying the chi square test.

SUGGESTIONS FOR FURTHER STUDY

A variety of lines of further investigation are possible; some suggestions follow. For background, begin by reading the review article by Elsner and Popov (1978).

1. Choice designs involving simultaneous playback from 2 speakers are a logical extension to this exercise. Females may show response to an acoustic stimulus presented alone which they would ignore in the presence of more preferred stimuli.

2. Temperature differences provide another line of investigation. Rates of forewing movement during stridulation are a linear function of temperature (Bessey and Bessey, 1898; Borland, 1972). Where the temperature of the tested female differs greatly from that of the recorded male, response will be much reduced.

3. Develop a variant of the experimental approach used in this exercise to investigate (a) the development (ontogeny) of phonotactic response in maturing female crickets, and/or (b) the effect of mating on subsequent sexual receptivity of females.

SUMMARY QUESTIONS

1. Are we justified in claiming from these experiments that sound is the basis of pair formation in crickets?

2. Design an experiment to test the possible role of male pheromones in attracting female crickets.

3. These experiments involved presentation of 1 stimulus at a time. How would our exploration of critical parameters (or releasers) be enhanced (altered?) by simultaneous presentation of 2 acoustic models?

4. Would you expect similar results if male or immature crickets were the test subjects instead of females?

SELECTED REFERENCES

Bell, P.D. 1979. Rearing the black-horned tree cricket, Oecanthus nigricornis (Orthoptera: Gryllidae). Can. Entomol. 111: 709-712.

Bessey, C.A. and E.A. Bessey. 1898. Further notes on thermometer crickets. Amer. Natur. 32: 263-264.

Borland, H. 1972. Cricket song. Audubon 74: 1.

Elsner, N. and A.V. Popov. 1978. Neuroethology of acoustic communication. Adv. Insect Physiol. 13: 229-355.

Morris, G.K., G.E. Kerr, and D.T. Gwynne. 1975. Calling song function in the bog katydid, Metrioptera sphagnorum (F. Walker) (Orthoptera: Tettigoniidae): female phonotaxis to normal and altered song. Z. Tierpsych. 37: 502-514.

Paul, R. 1976. Species specificity in the phonotaxis of female ground crickets (Orthoptera: Gryllicidae: Nemobiinae). Ann. Entomol. Soc. Amer. 69: 1007-1010.

Regen, J. 1913. Uber die Anlockung des Wiebchens von Gryllus campestris L. durch telephonisch ubertragene Stridulationslaute des Mannchens (Ein Beitrag zur Frage der Orientierung bei den Insekten). Pflüger's Archiv gesamte Physiologie Menschen und Tiere 155: 193-200. [Attraction of females of Gryllus campestris L. through telephone transmitted stridulation sounds of the males.]

Walker, T.J. 1957. Specificity in the response of female tree crickets (Orthoptera, Gryllicidae, Oecanthinae) to calling songs of the males. Ann. Entomol. Soc. Amer. 50: 626-636.

Ulgaraj, S.M. and T. J. Walker. 1973. Phonotaxis of crickets in flight: attraction of male and female crickets to male calling songs. Science 182: 1278-1279.

20. Defensive Behavior in Beetles

Janice R. Matthews
University of Georgia

Among arthropods, chemical weaponry is particularly diverse and well developed (Eisner, 1970; Blum, 1981). It ranges from very active defenses such as injected venoms manufactured in special glands through a wide range of "systemic" defenses found in blood, gut, or elsewhere in or on the body. These include urticating hairs, "rubbery" blood, enteric discharges, spines, tubercles, shields, and the like. Their effect is often amplified by behavior and morphology, as in aposematic coloration, reflex bleeding, and camouflage. Although these "systemic" defenses are usually thought of as passive, they overlap at the active end with actual forms of attack.

How effective are such types of defense? How do they act? In many cases, we can only guess; field observations of predator-prey encounters are few. From laboratory experiments, we know these defenses often act in multiple ways. The substances involved may not only be distasteful or malodorous, but may also act as topical irritants, systemic poison, mechanical entangling agents, or thermal agents.

This exercise limits itself to beetles--not a very great limitation when you consider that Coleoptera, the largest order of insects, includes 40% of the known species in the class Insecta. First, we will test the defenses of 3 common beetles, the mealworm, the ladybird, and the dermestid, against 2 aggressive potential predators--an invertebrate (fire ants, wood ants, or other species) and a vertebrate (chickens). Next we will attempt to develop a bioassay, using a known beetle defensive secretion in varying concentrations. Third we will investigate the survival of noxious insects and their mimics through trials using various chemically-altered and colored mealworms presented to chickens.

METHODS

Subject

Any of a large variety of beetles might be used as prey in these activities. Mealworm (Tenebrio molitor) larvae, ladybird beetles (also called "ladybugs"; any adult Coccinellidae), and dermestid beetle larvae are recommended. Available commercially,

135

they may be maintained in the laboratory. All should be chilled prior to use to slow their movement.

As invertebrate predators, laboratory colonies of fire ants (Solenopsis, see Exercise 16) or wood ants, Formica, have proven successful and are easily maintained. Other species (including other ants, salticid spiders, mantids, or other predators) might give interesting results. As vertebrate predators, chickens--easily obtained from farms or commercial growers--are lively and reliable performers. One per student or team, plus 1 or 2 extras to serve as "pinch-hitters", is recommended.

Materials Needed

Predators and prey as discussed above; feeding arenas for each chicken (an inverted laundry basket serves nicely); salicylaldehyde in various concentrations; distilled water; 66% aqueous solution of quinine dihydrochloride ($C_{20}H_{24}N_2O_2$ · 2HCl); micropipettes; forceps; microscope slides; filter paper; graph paper; various colors of permanent broad-point felt-tip markers, or quick-drying paints and camels' hair brushes; dissecting microscopes, preferably with at least one mounted on an adjustable counterbalanced horizontal arm for easy observation of the ant arena surface (see Fig. 16-1).

Procedure

A. Protection Against Invertebrate Predators

Obtain chilled mealworms, dermestid larvae, and ladybug beetles from their containers. For each species in turn, gently set an individual down on the ants' foraging platform closely adjacent to an active trail.

CAUTION: Do not bump the platform or breathe directly on it, as the ants are very sensitive to such disturbances.

You want to achieve as close to a "natural" discovery of a potential prey as possible. Observe what happens, using the low magnification of the the dissecting microscope if possible. At least 2 quantitative measures of the ants' response can be made: 1) time until the initial discovery or contact, and 2) time to overwhelm. Describe reactions both of the beetles and of the ants. Identify any defensive mechanism detected. Replicate at least 3 times for each prey species.

B. Protection Against Vertebrate Predators

Birds are generally acknowledged to be significant predators of insects, and certain aspects of defense such as aposematic coloration and mimicry may have developed primarily in response to vertebrate predation pressure. For the remainder of the exercise, we will use chickens as easily available, if not totally "natural" predators. Depending on the source of the chicken, there is a good possibility that it will also be "naive" as regards beetles.

As a control (for what?), we will use mealworm beetle larvae, which chickens normally relish. Place your chicken in the feeding arena and allow it to habituate to this environment for at least 5 min. Present a mealworm and note your chicken's behavior. At least 4 response categories are possible: not touch the larva, peck it, kill it, eat it. Repeat 5 times and record the chicken's reaction time for each trial (see the sample data sheet in Table 20-1). Does reaction time change with experience? Based on this, does refusal to touch a mealworm for 1 min seem ample as a criterion of rejection?

Introduce a chilled ladybug beetle into the feeding arena, and observe the bird's behavior. If the beetle is eaten, introduce a second one and repeat 5 times, recording reaction time for each trial as before. If a. given beetle stays uneaten for 5 min, remove it and introduce a mealworm. Discontinue trials if your chicken refuses to accept the mealworm.

Repeat using dermestid larvae and/or any other beetles provided. Graph the results of the pooled class data with response times (in sec) as the vertical axis and successive trials as the horizontal axis. Use of a different symbol for each species will permit all to be plotted on the same graph for ready comparison.

C. Developing a Bioassay for a Defensive Chemical

The effect of defensive secretions is usually immediate for both vertebrate and invertebrate predators. Variations of grooming behavior lasting seconds or minutes are the usual response, so it is not surprising that bioassays using grooming reflexes as a basis for determining "irritant effectiveness" have been developed. The time delay between application, of material and onset of scratching or other grooming maneuver is one useful measure of this (Eisner et al, # 1963).

A vial of salicylaldehyde is available. Smell it. This is the principal component of the defensive secretion of several different beetle species in at least 3 families (Eisner et al, 1963; Wallace and Blum, 1969, and references therein). Using a micropipette place a small drop of salicylaldehyde on a 1 cm^2 piece of filter paper and gently place it on the foraging table. Observe the ants' response.

Crush a small mealworm larva on a microscope slide, and divide it into 2 batches. Dab a very small amount of salicylaldehyde onto 1 batch with a micropipette, with the other serving as a control. Place the slide on the foraging arena, and compare the ants' response to the 2 baits. With some careful observation and thought, you should be able to identify 1 or more behavioral responses appropriate to measure "effectiveness" of the salicylaldehyde against ants--your own bioassay.

Once you have decided on an appropriate behavioral response to measure, obtain salicylaldehyde in varying concentrations and collect data for a series of bioassay trials recorded on a data sheet like the example in Table 20-1. A method for reliably delivering the same quantity of chemical to each assay unit will need to be developed, for only the differences in concentration are being assayed here. If time permits, replicate each assay concentration 3 times. Construct histograms of the pooled data plotting changes in number of ants responding to each of the treatments and controls over time.

TABLE 20-1. Example of a data sheet set up for recording chicken predation behavior (Part B), bioassay experiments (part C) and vertebrate learning (Part D) of this exercise.

Part B.

Beetle Tested		Chicken's Reaction Time (latency to response)(sec)				
	Trial No:	1	2	3	4	5
mealworm	1st peck:					
	eat:					
ladybird	1st peck:					
	eat:					
dermestid	1st peck:					
	eat:					
other	1st peck:					
	eat:					

Part C.

Salicylaldehyde Concentration	Number of Ants on Slide After:					
	1 minute		2 minutes		3 minutes	
	Trtmt	Cont	Trtmt	Cont	Trtmt	Cont
100%						
50%						
5%						
.5%						

Part D.

1. Odor Assay: Elapsed time till eaten (sec)

Trial No.	Chemical Trtmt	Water Trtmt	Control Trtmt
1			
.			
.			
10			

2. Avoidance based on color: Elapsed time till eaten (sec)

Trial No.	Painted Aposematic	Painted non-Aposematic	Unpainted Control
1			
.			
.			
10			

3. Avoidance Based on mimicry: Elapsed time till eaten(sec)

Trial No.	Models	Mimics	Controls
1			
.			
.			
10			

D. Chemical Defense and Learning in Vertebrate Predators

General Instructions: The class should divide the responsibility for the activities which follow. Each team will be offering different experimentally altered mealworms to their chickens. Use a table of random numbers (see Appendix 2), flips of a coin, or dice tosses to decide order of presentation of larvae, so that birds cannot learn to predict the order of various chemically altered and control mealworms. A particular chicken should be used in only one of the experiments. If your chicken refuses to eat the untreated control, disregard that trial. If this happens more than 3 consecutive times, discontinue trials with this bird. Prepare the chemically treated mealworm in an area separated from the experimental arenas to minimize possible bias in results. Pool class data as necessary, and split up the work of running appropriate statistical analyses.

1. Avoidance based on odor. NOTE: Chickens used in this activity should not be the same as those used in the salicylaldehyde bioassays above. Why?

Choose 20 similar sized mealworms. With a micropipette, dab salicylaldehyde on 5 of them; dab 5 others with an equal amount of distilled water. Place each mealworm individually on a fresh filter paper in a petri dish. (If mealworms are very active, you may need to crush their heads with forceps to immobilize them.)

Run 10 trials, using 2 dishes simultaneously (1 with a salicylaldehyde mealworm or distilled water mealworm, and 1 untreated control). For each trial slide the dishes into the cage, then move back to reduce human distraction. One person should time the chickens' response and another record. Be prepared for fast action. Randomize the order of presentation as described above, being sure to also vary right/left side positions of treated and control relative to the chicken. Thus, the chicken will have been presented with a total of 20 mealworms. If warranted, analyze your results by means of a 2 x 2 chi square test with Yates correction factor (see Appendix 1.1) to determine whether differences in the reactions of your bird to "smelly" inedible and "non-smelly" edible mealworms are real.

2. Avoidance based on color. With a felt marker or small brush and paint, mark a stripe of color down the back of several similar sized mealworms; let it dry thoroughly so it becomes odorless. These mealworms will constitute your "normal" non-aposematic sample.

With a different color, mark another set of mealworms. These will become your aposematic sample. (Red and orange are the most typical warning colors in nature; thus you might wish to use blue or another "unnatural" color to help control for the possibility that your chickens might have a previously learned aversion to the color.)

Dip your aposematic mealworms into a 66% solution of quinine dihydrochloride, an odorless substance unpalatable enough to cause other bird species to learn to reject mealworms on sight alone (Brower, 1960). Let them dry.

Following the procedure given in part 1 above, conduct 10 trials (total of 20 mealworms given to the chicken). Analyze your data with a 2 x 2 chi square test as above.

3. <u>Mimicry and Avoidance</u>. Prepare a set of banded mealworms following the procedure outlined above but using only a single color. Divide these mealworms into 2 groups. Dip one group into distilled water; these are your "mimics". Dip the other group into 66% aqueous solution of quinine dihydrochloride; these are your distasteful "models". (What type of mimicry situation are we duplicating, Batesian or Mullerian?)

Follow the presentation procedure outlined above. Each trial will consist of a randomized pair of prey offerings--one model <u>or</u> mimic, and one untreated edible. Record category of response and reaction time, plus any general notes on the bird's behavior. For example, does it show any displacement activities or signs of conflict when presented with a model or its virtually identical mimic?

SUGGESTIONS FOR FURTHER STUDY

1. Testing Protective Devices in Field-Collected Coleoptera

Anti-predator devices are of widespread occurrence, but most have not been tested experimentally. For those which have received attention, a great many novel defense mechanisms have been discovered (Table 20-2).

Depending upon seasonality and availability, obtain a variety of field-collected species of beetles. In Georgia in the past, we have successfully used larvae of these chrysomelids: <u>Gastrophysa cynae</u>, a common species whose larvae are restricted to dock (<u>Rumex</u>); the willow-feeding <u>Chrysomela</u> <u>scripta</u>; <u>Cassida</u> sp. feeding on morning glory; <u>Arthrochlamys</u> <u>plicata</u>, a case-bearer feeding on blackberry (<u>Rubus</u>). All but the <u>Cassida</u> were found only in spring. Table 20-2 offers other suggestions.

Begin by using low light intensity and the dissecting microscope to examine the external morphology of the species you have found. Using an insect pin probe, stimulate the beetle in an attempt to elicit defensive behavior. Touch various parts of the body. How fast is the response? How long does it last? Try repeated stimulation of the same spot. Now apply stimulation alternately on opposite sides in rapid succession and observe the dexterity of the response.

Present your beetle species to predators, and record behaviors as before. If you have chosen a shield- or case-bearing species, carefully remove the case or shield and expose the prey and the case separately to predators. Alternatively, take the encased larva and turn it on its side or back. What happens? Does your particular insect possess more than one defensive strategy? Do the results of these simple experiments tend to support the hypothesis of a defensive function of the various behaviors?

TABLE 20-2. Some common types of chemical defense in beetles. (Based on Eisner, 1970, wherein complete references are cited.)

Type of Defense	Examples	References
1. Glandular		
a. Eversible glands	Staphylinidae	Jenkins, 1957, & references therein
b. Oozing glands	Soldier beetles, (Chauliognathus)	Meinwald et al., 1968
	Cottonwood leaf beetle, (Chrysomela scripta) & other Chrysomelidae	Garb, 1915; Wallace and Blum, 1969
c. Spraying glands	Many Carabidae, incl. Galerita, Chlaenius, Calosoma, Helluomorphoides	Eisner, 1958; Moore & Wallbank, 1968; Aneshansley et al., 1969
d. "Reactor" gland	Certain Carabidae, incl. "bombardiers" (Brachinus), members of the Ozaenini, and Metrius	Eisner, 1958; Moore & Wallbank, 1968; Aneshansley et al., 1969
e. True poisons	Dytiscidae	Schildknecht et al., 1966
2. Non-glandular		
a. Blood & other systemic factors: reflex bleeding &/or easily ruptured veins, spines, etc.	Mexican bean beetle (Epilachna)	Happ & Eisner, 1961, 1966
b. Enteric discharges regurgitation, defecation; sometimes made into a "shield"	Chrysomelid beetle larva (Cassida)	Eisner et al., 1967
c. Detachable outgrowths and artificial coverings	Various dermestid beetle larvae	Nutting & Spangler, 1969

2. Chemical Effectiveness

The effectiveness of a defensive chemical probably depends on a great many factors, such as extent of contamination, sensitivity of sites affected, nature and concentration of the secretion, and type of predator. Design some experiments which would permit you to examine the role of one of these factors in detail. One possibility is to extract hemolymph of your coccinellid and test it directly on ants (see Tursch et al., 1971a, b).

3. Frequency of Mimics and Effectiveness of Mimicry

One of the basic tenets of Batesian mimicry has been that a model must occur in greater frequency than its mimic if the mimetic color pattern is to evolve and be maintained. However, when Brower (1960) undertook to experimentally investigate this question using 8 starlings as predators and painted mealworms as prey, she discovered that mimicry was effective even when the mimics were more frequent than models. How general and/or repeatable are her results? As a class, you might wish to investigate this question. Each team could offer their bird a different proportion of experimental and control mealworms for a series of 10 trials. Continue the trials over several days (Brower gave a total of 160 trials to each bird), and analyze your results statistically. Graphing the results is also useful. Percentage effectiveness of mimicry (number of mimics not eaten/total number of mimics, or number of mimics not touched/total) could be the vertical axis, with percentage of models or mimics on the horizontal axis. Are there real differences in the reactions of the individual birds to the various proportions of models and mimics? Are your results statistically different from those obtained by Brower?

4. The Perfect Defense

For fun, design your own "super beetle" equipped to defend itself against all contingencies. Ths instructor may wish to post entries and offer some appropriate award for the one judged best able to cope with the world of predators.

SUMMARY QUESTIONS

1. Characterize the defensive behaviors observed in part A.

2. Would you predict any of these beetles to be more vulnerable to another type of invertebrate predator? Another type of vertebrate predator? Explain.

3. Relate what you have observed to what you know of the ecology and life history of these beetle species. Can you make any generalizations correlating an insect's niche and possession of defensive devices of various types?

4. Does effectiveness of salicylaldehyde as measured by your bioassay appear to be concentration-related, or is it all-or-none once a certain threshold is crossed?

5. Based on the chickens' response to chemically treated mealworms, is there evidence of learning based solely on smell? On color? What category of learning might be occurring?

SELECTED REFERENCES

Blum, M. S. 1981. Chemical Defenses of Arthropods. Academic Press, New York.

Brower, J. van Zandt. 1960. Experimental studies of mimicry. IV. The reactions of starlings to different proportions of models and mimics. Amer. Nat. 94: 271-282.

Eisner, T. 1970. Chemical defense against predation in arthropods. In Chemical Ecology, E. Sondheimer and J. B. Simeone (eds.). Academic Press, New York, pp. 157-215.

Eisner, T., C. Swithenbank, and J. Meinwald. 1963. Defense mechanisms of arthropods. VIII. Secretion of salicylaldehyde by a carabid beetle. Ann. Entomol. Soc. Amer. 56: 37-41.

Nutting, W. L. and H. G. Spangler. 1969. The hastate setae of certain dermestid larvae: an entangling defense mechanism. Ann. Entomol. Soc. Amer. 62: 763-769.

Tursch, B., D. Daloze, M. Dupont, C. Hootele, M. Kaisin, J. M. Pasteels, and D. Zimmerman. 1971a. Coccinellin, the defensive alkaloid of the beetle Coccinella septempunctata. Chimea 25: 307-308.

Tursch, B., D. Daloze, M. Dupont, C. Hootele, M. Kaisin, J. M. Pasteels, and M. C. Tricot. 1971b. A defensive alkaloid in a carnivorous beetle. Experientia 27: 1380-1381.

Wallace, J. B. and M. S. Blum. 1969. Refined defensive mechanisms in Chrysomela scripta. Ann. Entomol. Soc. Amer. 62: 503-506.

21. Reproductive Behavior of Giant Water Bugs: I. Male Brooding

Robert L. Smith
University of Arizona

Male water bugs in the subfamily Belostomatinae carry eggs attached to their backs by conspecific females (Fig. 21-1). Prior to the 20th century, it was believed that egg-bearers were females carrying their own eggs. Early authors attributed the depositional process to a long protrusile ovipositor which the female bug was said to extend over her back. Credit for egg-carrying was properly bestowed on the male by a woman, Florence W. Slater, in an account which included the following:

> "That the male chafes under the burden is unmistakable; in fact, my suspicions as to the sex of the egg-carrier were first aroused by watching one in an aquarium, which was trying to free itself from its load of eggs, an exhibition of a lack of maternal interest not expected in a female carrying her own eggs."

Slater went on to say that when the male was attacked (by an unidentified agent), he "meekly received the blows, seemingly preferring death to the indignity of carrying and caring for the eggs." Another entomologist, in apparent accord with the "humiliated male hypothesis," stated:

> "The egg-bearing male . . . dislikes exceedingly this forced servitude, and does all he can to rid himself of the burden. From time to time he passes his third pair of legs over the dorsum, apparently in an endeavor to accomplish this purpose. If he is not able to get rid of it, as sometimes happens, he carries his burden till in due time all the little ones are emerged, when he at last frees himself from it."

It should be clear that these interpretations of observed behavior are not compatible with modern selection theory. Natural selection could not have favored females programmed to dispose of their eggs on the back of a male only to have them discarded in places or conditions that might impede their development. On the contrary, females are always under intense selection to choose oviposition sites that will maximize egg viability. Therefore, the back of a male should be the optimal oviposition substrate for species belonging to the belostomatid subfamily characterized by this behavior. The behavior of egg-bearing (encumbered) males should somehow maximize egg viability. The purpose of this exercise is to discover the function of male egg carrying, and to reconcile the observations and misinterpretations of early workers.

METHODS

<u>Subject</u>

Giant water bugs (Hemiptera: Belostomatidae) are large, brown, oval insects with raptorial front legs. In ponds and lakes, they feed on other insects, snails, tadpoles, and small fish. If handled carelessly, they are capable of biting.

Three genera of giant water bugs occur in the United States -- <u>Lethocerus</u>, <u>Belostoma</u>, and <u>Abedus</u>. <u>Abedus</u>, long-lived and easily reared and maintained in the laboratory, is recommended for these exercises (see Aids to the Instructor).

Fig. 21-1. Female giant water bug, <u>Abedus herberti</u>, laying eggs upon her mate's back, where he will brood them until hatching.

Materials Needed

For each team of students: 4 egg-encumbered male and 1 female belostomatid (Abedus herberti or A. indentatus); 5 one-liter plastic cups or similar vessels (such as 1 qt ice cream or cottage cheese containers), with lids; 2 two-liter glass aquaria; aquatic plants (such as Elodea) and some rough angular stones (3-5 cm diam); stopwatch or event recorder; distilled water; scalpel or spatula.

Three first instar nymphs will be needed by each team 2 weeks after the experiment is begun.

Procedure

A. Factors Influencing Water Bug Hatching Success

Remove the egg pad from 1 of the encumbered males by gently lifting the anterior edge with a small spatula or the dull edge of a scalpel. Working from anterior to posterior, peel the intact egg pad from the male's back. Divide the pad into 3 portions, each containing approximately equal numbers of eggs. Place 1 portion, attached side down, in each of 3 plastic cups. In 1 cup add distilled water to a depth of about 4 cm. In a second cup place just enough water to moisten the eggs, such that the unattached ends are exposed to air. Leave the third cup dry. Cover all 3 with perforated lids to reduce evaporation. Place several stones in the 2 remaining cups. Add water to a depth of about 10 cm above the stones in 1 cup, and to about 1 cm above the stones in the second cup. Place an encumbered male in both cups, and cover each with a perforated lid. Set all the cups aside and observe all treatments twice each week. Record the condition of eggs and any significant events on the data sheet (Table 21-1).

Discuss your results. If eggs are observed hatching, describe the hatching process, the behavior of the hatching nymphs, and their behavior after eclosion.

B. Behavior of Brooding and Non-brooding Males

Set up 2 aquaria to be as nearly identical as possible, with distilled water to a depth of about 10 cm, rocks, and floating aquatic plants. In 1 aquarium place an encumbered male. In the other, place the male from which you removed the eggs used in the previous experiment. Allow about 30 min for the bugs to acclimate, then alternately observe the encumbered and the unencumbered male for 10 min each. Be especially alert for subtle differences in behavior between the brooding and nonbrooding bugs. Compare your observations with those of other observers in the class.

List behavior patterns observed for encumbered and unencumbered males and note any qualitative differences between them. Design and carry out timed observations or event recordings to detect quantitative differences in behavior between brooding and non-brooding males.

TABLE 21-1. Data sheet for reproductive behavior of giant water bugs.

1. EFFECTS OF MALE BROODING ON EGG VIABILITY

Experimental Treatment	Week 1 date____	Week 2 ____	Week 3 ____	Week 4 ____
DRY EGGS				
MOISTENED EGGS				
SUBMERSED EGGS				
EGGS ON MALE IN SHALLOW WATER				
EGGS ON MALE IN DEEP WATER				

2. HATCHING BEHAVIOR:
   Date:_____ Description:

3. NYMPHAL BEHAVIOR AFTER ECLOSION:
   Date:_____ Description:

C. Cannibalism and Brooding Behavior

Deprive a brooding and a non-brooding adult male and a female of food for 2 weeks. Present each of the "hungry" individuals with a first instar conspecific nymph. Note whether or not the nymph is cannibalized. Pool data from the entire class and determine whether there are significant differences in cannibalistic proclivity among females and brooding and non-brooding males. Continue to starve your brooding male and observe his reaction to first instar nymphs that hatch from his back.

SUMMARY QUESTIONS

1. What conclusions can you draw concerning the "needs" of _Abedus_ eggs and emerging nymphs?

2. Could eggs submersed in still water hatch unaided by a brooding male?

3. How would you characterize male brooding? What are its behavioral components?

4. How do male brooding patterns contribute to egg survival?

5. What costs might be incurred by males as a consequence of brooding?

6. What do the results of your nymph cannibalism experiment have to do with brooding?

7. What do you suppose is the genetic relationship of brooding males to the eggs they brood?

8. Members of the subfamily ancestral to brooding belostomatids (the Lethocerinae; Lauck and Menke, 1961) lay their eggs on emergent vegetation above the surface of the water. Neither sex broods. How might male brooding in the subfamily Belostomatinae have evolved from the lethocerine system?

SELECTED REFERENCES

Cullen, M.J. 1969. The biology of giant water bugs (Hemiptera: Belostomatidae) in Trinidad. Proc. R. Entomol. Soc. Lond. (A)44: 123-137.
Lauck, D.R. and A.S. Menke. 1961. The higher classification of the Belostomatidae (Hemiptera). Ann. Entomol. Soc. Amer. 54: 644-657.
Smith, R.L. 1976. Male brooding behavior of the water bug _Abedus herberti_ (Hemiptera: Belostomatidae). Ann. Entomol. Soc. Amer. 69: 740-747.
Smith, R.L. 1976. Brooding behavior of a male water bug, _Belostoma flumineum_ (Hemiptera: Belostomatidae). J. Kans. Entomol. Soc. 49: 333-343.

22. Reproductive Behavior of Giant Water Bugs: II. Courtship, Role Reversal, and Paternity Assurance

Robert L. Smith
University of Arizona

Male giant water bugs in the subfamily Belostomatinae invest time and energy brooding eggs attached to their backs by conspecific females (see Exercise 21). Natural selection theory predicts that a male should care only for young that possess his genes. Males should certainly avoid being "cuckolded" (Trivers, 1972) -- that is, being tricked into caring for embryos or young that are the product of sperm from another male. Most male insects contribute nothing to reproduction beyond sperm, and are therefore not threatened with cuckoldry. The water bugs that brood eggs are threatened, and should be expected to possess refined anti-cuckoldry or paternity assurance adaptations.

Another evolutionary prediction that arises out of the brooding behavior of male water bugs concerns the alteration of roles in courtship. Theory forecasts that whichever sex contributes most to reproduction becomes a limited resource, and should be courted by the other sex. In the overwhelming majority of animal species, the female makes the larger investment, and hence females are usually courted by males. If male water bugs make the larger investment in reproduction, then females should be expected to court them.

This laboratory exercise has been designed to permit you to observe, describe, and analyze a complex courtship and mating behavioral sequence and to test predictions concerning paternity assurance mechanisms and reversal of roles in courtship when males make a significant contribution to reproduction.

METHODS

Subject

This laboratory exercise utilizes giant water bugs (Hemiptera: Belostomatidae: Belostomatinae), preferably <u>Abedus</u> <u>herberti</u> or <u>A.</u> <u>indentatus</u>. For further information, see Exercise 21 and the Aids to the Instructor.

Materials Needed

For each team: 1 male and 1 gravid female giant water bug; 2-liter aquarium; stopwatch; tape recorder; hydrophone (or harmonica contact microphone sealed with silicon rubber), amplifier, and earphones; several rough angular stones (3-5 cm diam); distilled water; liquid typewriter correction fluid; tissues or laboratory wipes.

Procedure

A. Observing the Courtship and Mating Behavioral Sequence

Remove the gravid female from her holding container, and blot her dorsum dry with a tissue. Place a spot of typewriter correction fluid on her scutellum and allow it to dry before placing her in the aquarium. Place male and gravid female together in the aquarium containing a few stones and distilled water to a depth of 5-8 cm. (Bugs may be sexed by examination of the subgenital plate, Fig. 22-1).

NOTE: Observers should stand at least ½ meter away from the aquarium, and avoid any abrupt movement, particularly any that would cast a shadow over the surface of the water. Frightened bugs take up to 20 min to settle down.

Immediately after the bugs are placed together in the aquarium, start the stopwatch and be prepared to record events as they take place. Use the "pause" button on your tape recorder to conserve tape when nothing is happening. Real elapsed time should be read from the stopwatch frequently after the description of events. If your bugs are "behaving well," continue to record their behavior for 60-90 min. Now produce an inventory of discrete behavior patterns observed, e.g., male approaches female; female strokes male with hind legs; male grasps female with foreleg; copulation, etc. Use your inventory and tape recording to diagram sequences of events on a time scale (you may wish to use symbols for each behavior in your inventory). Discuss any patterns that emerge. Relate your findings to Fig. 22-2.

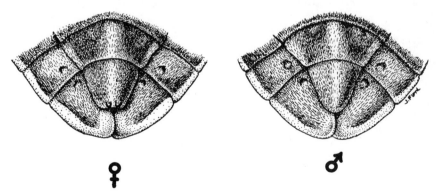

Fig. 22-1. In ventral view, males can be distinguished easily from females by their broadly pointed genital plates. The female genital plate is blunt, and ornamented with two tufts of fine hair.

B. The Role of Sound in Abedus courtship

Arrange stones in the bottom of an aquarium so they form an "island" covering your hydrophone or contact microphone. Now add water to a depth of about 5 cm over the pile of stones. Plug the microphone into your audio amplifier and adjust to a comfortable volume. You should be able to hear a loud sound when one of the rocks is brushed lightly with a finger. Add a male and gravid female bug, and begin observing and listening. Keep a record of all events including sounds produced by the bugs. You can make crude recordings of sounds by holding the tape recorder microphone next to a speaker or earphone. Recordings may be replayed for casual analysis. Characterize the sounds heard and place sound production in its behavioral context. Speculate on its function. Do isolated males and females produce sounds? Do both sexes produce sounds? If so, which sex is the more "vocal?"

SUMMARY QUESTIONS

1. How are roles in courtship reversed for these animals?

2. Do males possess behavior patterns that may permit them to assure paternity of eggs they receive to brood? If so, how does the system work? How might it have evolved?

3. What male pattern "tells" the female something about the individual's ability to brood?

4. What kind of selection probably caused the evolution of the display referred to above (3)?

5. Which sex controls the sequencing of mating and oviposition?

6. How many times could the bugs copulate in 24 hours? Extrapolate from your short-term data.

7. What may be the function of sounds produced in the context of courtship and mating?

SELECTED REFERENCES

Parker, G. A. 1970. Sperm competition and its evolutionary conse-
    quences in the insects. Biol. Rev. 45: 525-567.
Smith, R. L. 1979. Paternity assurance and altered roles in the
    mating behaviour of a giant water bug, Abedus herberti (Heterop-
    tera:Belostomatidae). Anim. Behav. 27: 716-725.
Smith, R. L. 1979. Repeated copulation and sperm procedence: Pater-
    nity assurance for a male brooding water bug. Science 205:
    1029-1031.
Smith, R. L. 1980. Daddy water bugs. Natural History 89: 56-63.
Smith, R. L. 1980. Evolution of exclusive postcopulatory paternal
    care in the insects. Fla. Entomol. 63: 65-78.

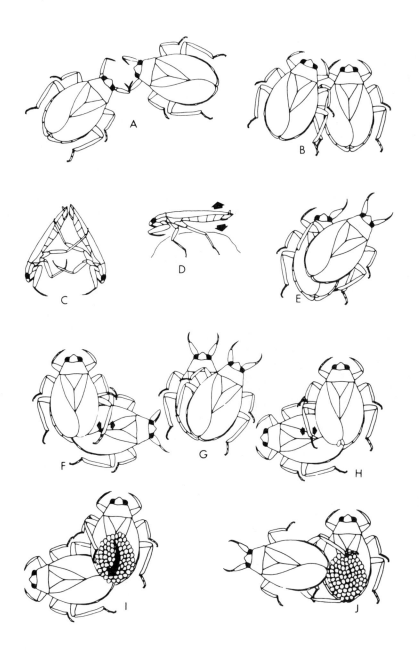

Fig. 22-2. Courtship sequence of giant water bugs. What is happening in each step pictured here?

23. Sexual Behavior of Damselflies

John Alcock
and Ronald L. Rutowski
Arizona State University

The sexual behavior of any species of animal involves a complex relationship between the behavior of males and females. A modern understanding of the adaptive features of sexual behavior begins with the observation that males produce far more sperm than females produce eggs. This, coupled with the observation that the number of males usually exceeds the number of <u>receptive</u> females in a population, suggests that females and the eggs they contain are a limited resource for males. In species such as damselflies in which males make no investment in the care of their offspring, a male's ability to propagate his genes will be a function of the number of eggs he fertilizes. In these species, natural selection should favor males that maximize their opportunities to meet, court, and copulate with females. Males should search for mates at times and in places where receptive females are most likely to to be encountered. They should possess the ability to recognize females of their species accurately and rapidly. And they should monopolize access to females if they can do so efficiently.

A male's genetic success may be influenced not only by his ability to detect and copulate with females but also by whether or not he guards a recent mating partner. In some damselflies, a female may copulate with several males but only use sperm from the most recent mating to fertilize her eggs. This phenomenon, known as "sperm precedence", means that a male may gain nothing from a copulation if his mate accepts another male before laying her eggs. Thus, males may gain by defending their mates against competitor males under certain circumstances.

The object of this exercise is to examine the nature of sexual interactions in damselflies under natural conditions, especially aspects of mate location and recognition by males. Through behavioral observation, censusing, and experimental manipulation we will attempt to discover whether males behave in a way that maximizes their chances of transmitting genes in competition with other males.

METHODS

Subject

Damselflies are common, often abundant, insects along the marshy borders of ponds, lakes, and slow-moving streams from spring to early

154

fall throughout North America. Because they are generally more numerous than dragonflies and can be approached more closely, they make better subjects for a laboratory exercise than their larger relatives. Nonetheless, the questions posed here about damselfly behavior can equally be applied to dragonflies in areas with large populations of these insects.

The most rewarding study area will be one with a relatively dense population of the selected species and with an accessible water's edge. The species selected for study should be one in which both males and females can be easily and reliably distinguished. Damselflies often have color markings that are species and sex specific but change with age. Newly emerged individuals are generally of a duller color than mature adults; in addition, they are soft-winged and relatively inactive. By focusing upon pairs in tandem (Fig. 23-1), one can easily learn to recognize the distinguishing markings of a mature male and female of the same species.

Materials Needed

Insect nets; flagging materials for marking census transect intervals; meter tape; stop watches or wrist watches with second hand; 15 cm ruler; 1 to 2 m "fishing pole" with thread tied to end; several colors of acrylic or enamel paints (such as Testor model paints) or liquid typewriter correction fluid (such as Liquid Paper); data sheets; mapping paper; centigrade thermometer.

Procedure

You may wish to work individually or in pairs with 1 person to observe and the other to record and time the behaviors under study. Begin by observing tandem pairs until you can reliably identify conspecific individuals of both sexes. Keep careful and precise records. For several of the parts which follow, data from the entire class will be compiled near the end of the exercise.

A. The Components of Damselfly Reproductive Behavior

To begin studying the male-female interactions leading to copulation, follow the activities of individual males or groups of males. Notice any rapid pursuit and capture of 1 individual by another. Observe the activities of pairs already copulating. The gonads of males and females are near the tips of their abdomens, but members of a pair do not link abdominal tips during mating. How is sperm transfer conducted?

From such casual observations, it should be apparent that males do not respond in the same ways to conspecific females and to other damselflies but that they vary their response to different classes of damselflies. Do males reliably recognize females of their own species? Quantitative documentation is needed to answer the question.

Male damselflies commonly respond to any other passing damselfly by: approach -- male flies to within 5 cm of another individual;

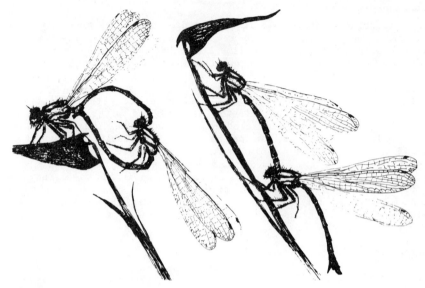

Fig. 23-1. Mating and guarding in a damselfly. Left: the female bends the tip of her abdomen forward to contact the male's accessory genitalia located on the underside of his second abdominal segment. The male uses claspers at his abdominal tip to grasp the female's prothorax, thus completing the "wheel position" characteristic of this group. Right: a pair resting on vegetation with the male still clasping the female in the tandem position as she prepares to oviposit. (Drawing by J. W. Krispyn).

chase -- male follows another individual in flight for more than 3 sec; hover -- male remains motionless in space before a perched damselfly for more than 3 sec.

Construct a data sheet and collect a random sample of observations starting with the first male you encounter. Record the sex and species of any other damselfly that passes within 10 cm of the selected male. Note the male's response or lack of reaction. Terminate your observation of that individual after its first reaction or after a standard period of time (a few minutes). Pick another male at random and repeat the procedure. If each student makes 10 to 20 records and the class pools their data, these should be sufficient to test the hypothesis that males react differently to females of their species than to conspecific males, foreign females, or foreign males.

B. The Nature of Visual Releasers

What cues does a male use to identify females? If your species is sexually dimorphic in color pattern, a likely possibility is color cues.

Collect some female damselflies in an insect net and kill them by crushing the thorax. Tie a female to a "fishing line" (a piece of thread tied to a thin wooden rod). Dangle or move the specimen in front of a male. Will he respond to a freshly killed female? If so, you can alter the color patterns on the body and/or wings of collected damselflies with paints. Make females resemble males more or less closely; take some male specimens and paint them to look like females to a greater or lesser degree. Systematically present your altered specimens to males and record their responses as before.

If time and population size permit, you may wish to meet with your instructor and class members to determine what seem to be the best set of altered specimens with which to test more rigorously the role of color cues in mate recognition. Make up a standard set of specimens and collect data which can be pooled to create a large sample to test your hypothesis on the nature of the visual releasers of male sexual behavior in your damselfly species.

C. Activity Patterns

Typically, female damselflies feed upon small insect prey some distance from water and only come to the pond or stream when they have fed sufficiently to produce a clutch of eggs to be laid in aquatic vegetation. It is at this time that they are receptive to males. Males therefore tend to concentrate their mate-location activities in the vicinity of oviposition sites. Females rarely oviposit evenly throughout the day. Reportedly they are strongly influenced by cyclic differences in air temperature. We would predict that males should also focus their searching activity in a way that would take this factor into account.

To test this possibility, you will need to collect information on the pattern of daily activity in your species. As a class, select a strip of stream or pond edge along which to set up a transect of appropriate length (100-200 m) subdivided into 5-10 smaller units. Divide the class into census teams or individuals responsible for different portions of the day. At suitable intervals (30-60 min), a team or individual should walk along the census strip, stopping at the determined intervals to record the information listed on the sample data sheet (Table 23-1).

Later in this exercise, we will examine these data further. At this point, however, you should be able to use them to answer such questions as (1) What is the observed sex ratio? Does it change throughout the day? (2) Do females tend to come and copulate and/or oviposit more often at certain times of the day than at others? What relationship to temperature does this have, if any? (3) Does male searching activity change over the day? Do more males tend to be present and/or more active at the optimal time to encounter mates?

D. Male-Male Interactions

Damselflies range from species with completely non-aggressive patrolling males to species with highly aggressive territorial males. Why does such a spectrum exist?

Evolutionary theory suggests that territorial behavior evolves when the reproductive gains derived from excluding competitors in an

TABLE 23-1. Damselfly census data from transect.

Observer:		Time:			Temperature:
	No. of Individuals at Transect Checkpoint No.:				TOTAL
	$\underline{1}$	$\underline{2}$	$\underline{3}$	$\underline{4}$ ... $\underline{n}$	
Single Males:					
Flying					
Perched:					
"alone"					
nr. ovipositing					
female Other:					
Single Females:					
Flying					
Perched					
Ovipositing					
Other:					
Pairs					
Copulating					
Tandem Perched					
Tandem Ovipositing					
Other:					

area outweigh the costs of territorial defense. Such costs usually involve the time and energy expended in repelling intruders, expenses that reduce chances to find females and physically drain the territorial defender, shortening his life. Territorial defense becomes more expensive with increases in the rate at which intruders invade a territory, requiring the owner to respond. But territoriality becomes increasingly valuable to the degree to which oviposition sites are small, compact patches separated by other less suitable areas. When superior oviposition sites are few and are readily identified by males and females alike, a male that can claim a site as his own will be able to monopolise a disproportionate share of receptive females. When oviposition sites are more or less uniformly distributed over a broad area (and/or are not readily identified), defense of a territory will fail to reap such a large reproductive reward.

From your pooled class census data, you can determine both (1) the density of males and/or frequency of encounters between males, and (2) the degree of clumping of oviposition sites.

You may also have sufficient data on male-male encounters from parts A and C. If you wish, male-male encounters may be measured further by selecting a male at random and counting the number of intrasexual interactions in 5 or 10 min observation periods distributed evenly over the peak period for mating. The clumping of oviposition sites can be established further by drawing a map of the study area and recording the location of every ovipositing female seen in sample intervals spread over the peak period for oviposition.

Ideally these measures would be collected for 2 different species, 1 of which was territorial and the other not. Perhaps your instructor can provide comparative data from the literature or from previous class studies. But even without such information, you should be able to gain at least a general impression about the relative costs of territoriality vs. non-territorial searching, and how evenly or unevenly oviposition areas are distributed.

If your species seems to exhibit territoriality, you may wish to mark individual males with paints on the abdomen, wings and/or thorax after catching them in insect nets from their perches. Although delicate in appearance, damselflies can accommodate considerable painting and handling.

When you release a marked male, does he return to his perch at once? Keep track of the movements, if any, of marked males over a period of time (hours, or even days). How long can a territorial male hold his patch? How large an area does he claim? Do territories appear to enclose superior oviposition sites? Do some territories appear to attract larger-than-average numbers of females?

E. Mate Guarding

Do males prevent their mates from copulating again for some time after mating with them? If we assume that damselflies are "typical" insects with regard to sperm precedence, we can predict that male damselflies should guard their mates (1) if their partners are likely to copulate again if left unguarded, and (2) if there are relatively few opportunities for the males to mate repeatedly in the course of a day. While protecting his mate, a male's chances of locating and copulating with additional females are presumably reduced. When there are few receptive females available during a mating period, mate guarding costs are reduced because a male is unlikely to miss a potential chance to copulate while stationed by his previous mate.

Male damselflies that guard their mates do so either by remaining perched near the ovipositing female or by remaining in tandem while the female oviposits. If your species appears to engage in guarding, you can test the likelihood that females will remate if left unguarded. Remove the guarding male and observe what happens. (This may involve physically separating a tandem pair and releasing the female.) Perform a series of these simple experiments. What is the average length of time to remating for the female?

If your species exhibits mate guarding, you may wish also to undertake further studies to test the prediction that the probability of repeated copulation is low for a male. Follow mating pairs from the time of copulation to separation to gain some idea of how long males invest in guarding a mate. Then measure the rate of arrival of receptive females to a census transect relative to the number of males present at various times. If the male to arriving female ratio is high, then the probability of mating by any 1 male is low, and the cost of guarding, in terms of missed opportunities to mate, is also low.

If males of your species apparently fail to guard, is there a high proportion of receptive females to searching males? Does a mated female disappear completely underwater to oviposit, thus making it impossible for her to mate again, at least while she remains submerged?

F. Concluding Thoughts

Because the study of damselfly behavior will in most cases be done with a single species, your analysis of the adaptive value of elements of the insect's behavior will largely take the form of demonstrating that there is a plausible relationship between an environmental factor that influences individual reproductive success and a behavioral trait. For example, because females of some species will mate again readily if left unguarded (probably reducing the genetic gain to males that copulate and leave their mates), it may be adaptive for males of some species to remain in tandem with their mates while they oviposit.

This sort of argument is reasonable, but less persuasive than it would be with supporting information from studies of the males of other species that also guard their mates. If females of these species also mate again quickly after release by a male and if males suffer a genetic loss as a result, one would expect that these males would also guard their mates while they lay eggs. Studies with evidence supporting such hypotheses are often found in the literature, and should be cited in any discussion of the behavior of the particular species you have studied. Under Selected References below, a series of papers on dragonfly and damselfly behavior are listed that are likely to contain such information. Similar information on more distantly related organisms may also be gleaned from the literature and included in your report.

SUMMARY QUESTIONS

1. Summarize the nature of sexual interaction in damselflies. How do males and females of 1 species locate one another? How do females acquire sperm from their mates?

2. Summarize what you know of the pattern of daily activity of your species. Is it the same for both sexes? If not, how does it differ? What might be the significance of this difference?

3. How do males maximize their chance of encounter with receptive females?

4. What is the possible significance of male territoriality? Of mate guarding by males?

SELECTED REFERENCES

Alcock, J. 1979. Multiple mating in Calopteryx maculata (Odonata: Calopterygidae) and the advantage of non-contact guarding by males. J. Nat. Hist. 13: 439-446.

Bick, G.H. and J.C. Bick. 1965. Demography and behavior of the damselfly Argia apicalis (Say) (Odonata:Coenagriidae). Ecology 46: 461-472.

Bick, G.H. and L.E. Hornuff. 1966. Reproduction behavior in the damselflies Enallagma aspersum (Hagen) and Enallagma exulsan (Hagen) (Odonata:Coenagriidae). Proc. Entomol. Soc. Wash. 68: 78-85.

Campanella, P.J. and L.L. Wolf. 1974. Temporal leks as a mating system in a temperate zone dragonfly (Odonata:Anisoptera). I. Plathemis lydia (Drury). Behaviour 51: 49-87.

Jacobs, M.E. 1955. Studies on territorialism and sexual selection in dragonflies. Ecology 36: 566-586.

Johnson, C. 1962. A description of territorial behavior and a quantitative study of its function in males of Hetaerina americana (Fabricius) (Odonata:Agriidae). Can. Entomol. 94: 178-190.

Parker, G.A. 1970. Sperm competition and its evolutionary consequences in the insects. Biol. Rev. 45: 525-567.

Paulsen, D.R. 1974. Reproductive isolation in damselflies. Syst. Zool. 23: 40-49.

Waage, J.K. 1973. Reproductive behavior and its relation to territoriality in Calopteryx maculata (Beauvois) (Odonata:Calopterygidae). Behaviour 47: 240-256.

Waage, J.K. 1979. Dual function of the damselfly penis: sperm removal and transfer. Science 203: 916-918.

24. Courtship of *Melittobia* Wasps

Robert W. Matthews
University of Georgia

Courtship among insects is highly diverse, ranging from simple mounting without preliminaries to elaborate stereotyped sequences. Typically, complex courtship will include behaviors of attraction, recognition, orientation, mounting, antennation, and finally copulation. Visual, chemical, and tactile cues are almost always involved. The actions of the sexes are highly interdependent. In complex and delicate coordination, each responds to the behavior of the other and, by its own response, triggers (or "releases" in the language of the classical ethologists) the next step in the sequence. Such interactions are often schematically represented by a "reaction chain", a diagram which indicates both specific behaviors and the order in which they occur (see Fig. 1-1).

Reaction chains can be quite helpful tools for visualizing many temporal sequences of arthropod behavior which tend to be quite stereotyped. However, they have received their widest usage in analyses of insect courtship. In this exercise, we will characterize the nature of courtship in Melittobia wasps in terms of behavioral reaction chains, as a means of analyzing the elements of a complex behavior sequence. We will attempt to determine the sensory modalities involved, the role of male-male and female-female interactions in courtship, and the correlation of male and female antennal structure to courtship role.

METHODS

Subject

Melittobia is a tiny (slightly smaller than Drosophila), harmless chalcid wasp belonging to the family Eulophidae. A gregariously developing parasite upon the prepupae of a wide variety of insects (Fig. 24-1), it is easily obtained from mud dauber wasp nests and can be cultured upon blow fly (Sarcophaga) puparia, which may be purchased from biological supply companies (See Aids to the Instructor).

Observations of courtship in Melittobia are facilitated by the pronounced sexual dimorphism encountered in this genus. Males are brown, blind, short-winged, and possess bizarrely modified antennae, whereas the normal-winged and sighted black females do not (Fig.24-2).

Fig. 24-1. Immature stages of <u>Melittobia</u> developing on <u>Trypoxylon</u> host. Left: mass of gregarious prepupae; the fully fed larvae overwinter in this stage, prior to voiding their gut contents, (note the strings of feces), and metamorphosing into pupae (right).

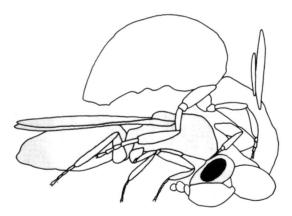

Fig. 24-2. A pair of <u>Melittobia</u> wasps engaged in courtship. The male (above) is blind and unable to fly. The bases of his antennae are greatly swollen and used to clasp the female's antennae during courtship. (From Evans and Matthews 1976)

Materials Needed

Wasp cultures; observation chambers (see Aids to the Instructor for construction hints); rectangular glass cover slips; dissecting microscopes and lamps fitted with heat filters; camel's hair brushes; watches with second hands; filter paper.

Procedure

Prior to the class meeting the instructor will have isolated virgin females and males. Be certain you have the microscope and lamp set up and focused before placing the sexes together in the observation chamber.

Work in pairs, with 1 person to observe and the other to record and time the various components of a normal courtship sequence. At the termination of each courtship, the 2 persons should switch roles. At least 1 of the pair should possess a watch with a second hand.

For each trial, first transfer the male, using the camel's hair brush. Observe briefly. Note the unusual antennae and the absence of eyes. Then as carefully as possible introduce the female and begin observing any interactions. Note: do not worry if occasional females escape or are injured in the process of learning this skill, since virgin females should be in ample supply. A brief period of exploration and acclimatization to the chamber may be necessary before sexual activity is initiated. If no mating behavior is observed within 5 min score the pair as "no mating" and go on to another pair.

Begin timing when the sexes make initial tactile contact. With the first pair, attempt to identify the individual behavioral components of the courtship interaction. Set up a data sheet with a column listing male behaviors and another for those of the female. Attempt to identify and label discrete behavioral components as objectively and precisely as possible -- for example, front wing raising, antennation, middle leg lift, etc. Some components will be repeated; attempt to identify any such characteristic behavioral sequences. By the second and third trials you should have a fairly good "feel" for the overall sequence of behavioral acts, and be able to construct a tentative flow chart (reaction chain) for the court-ship. As you become familiar with Melittobia courtship, attempt to obtain more refined quantitative data on the duration of each behavioral component and the frequency of repeated acts.

After a successful mating, transfer the individuals to holding containers. Since males may be scarce, they may need to be used more than once, but be sensitive to the possibility of changes in male behavior as a result of experience.

SUGGESTIONS FOR FURTHER STUDY

1. The sex ratio in Melittobia has been reported to be approximately 1 male: 50 female (Evans and Matthews, 1976). What might you post-ulate concerning the male reproductive strategy? The female? Design an experiment to determine whether either sex can mate more than once. Are there any detectable differences in the behavior of "experienced"

versus "inexperienced" individuals?

2. In the literature there are conflicting reports concerning the occurrence of Melittobia male aggression. Males have been seen to kill male nestmates on occasion. Design an experiment to examine male-male interactionˢ. Carry out the experiment and record behaviors observed. Matthews (1975) provides some additional background information on agonistic behavior in Melittobia. Additional experiments to investigate the roles of variables such as male size, age, and experience in aggressive behavior might prove worthwhile.

3. A very crude bioassay for the presence of an insect pheromone is provided by simply squashing the insect upon a piece of filter paper, then offering the paper to conspecific individuals. Prepare such squashes for both male and female (virgin) Melittobia. (Should you also have a control? If so, what?) For each trial, record the response (as percentage of time spent investigating the squash) for a virgin female or male over a 3 min interval in the observation chamber or a small covered petri dish. Replicate at least 4 times with different individuals and analyze your data. Would you expect differing responses from virgin and non-virgin females? Males? Why or why not? Design experiments using the bioassay technique to test your predictions (see Hermann et al., 1974).

4. Investigate aspects of the next stage of reproduction -- oviposition -- using mated females from your study. Following introduction of females into containers with Sarcophaga puparia (see Aids to the Instructor), observe behaviors associated with host finding and oviposition. Variations might be to investigate female competition for a single host, or the changes in a female's behavior with presentation of successive hosts. Two females could be recognized individually by carefully wing-clipping one.

5. The parasitic wasps offer unusual opportunity for undertaking comparative studies of courtship. As a start try a related chalcid, Nasonia vitripennis, readily obtainable from biological supply houses and also cultured on Sarcophaga. Barash (1975) provides a detailed discussion of procedures and results using Nasonia.

SUMMARY QUESTIONS

1. How would you characterize the nature of Melittobia courtship in terms of a behavioral reaction chain? What sensory modalities are involved? What is the relative timing of the various steps? How variable is courtship in this wasp, qualitatively and quantitatively?

2. What correlations can you draw between male and female antennal structure in Melittobia and courtship behaviors? Is there any evidence for pheromonal usage? How might you design further relevant experiments on this topic? What would you expect them to show?

3. What selective advantages and disadvantages might there be to

such strong reliance upon stereotyped sequential interactions so often found in mating behaviors?

4. What factors may have been responsible for the evolution of such strong sexual dimorphism in this genus?

5. What roles might male-male and female-female interactions play in Melittobia courtship?

SELECTED REFERENCES

Barash, D.P. 1975. Reaction chains in insect courtship: Nasonia vitripennis. In Animal Behavior in Laboratory and Field, E. O. Price and A. W.Stokes (eds.). W.H. Freeman, San Francisco. pp. 64-66.

Evans, D.A. and R.W. Matthews. 1976. Comparative courtship behavior in two species of the parasitic chalcid wasp Melittobia (Hymenoptera:Eulophidae). Anim. Behav. 24: 46-51.

Hermann, L.D., H.R. Hermann, and R.W. Matthews. 1974. A possible calling pheromone in Melittobia chalybii (Hymenoptera: Eulophidae). J. Georgia Entomol. Soc. 9: 17.

Matthews, R.W. 1975. Courtship in parasitic wasps. In Evolutionary Strategies of Parasitic Insects and Mites, P. W. Price (ed.). Plenum Press, New York. pp. 66-86.

25. Food-Based Territoriality and Sex Discrimination in the Water Strider *Gerris*

R. Stimson Wilcox
S.U.N.Y., Binghamton

Water striders (Family Gerridae) may be found on the surface of practically any temperate or tropical pond, river, or ocean in the world. These graceful insects are unusually easy to find, observe, and maintain, because they live mainly in 2-dimensional open water areas, will live on any clean water surface, and will feed on almost an prey small enough for them to subdue. There are many strider species in both streams and ponds upon which a variety of projects could be done, including, of course, comparative studies. Furthermore, there are other surface-dwelling groups (e.g. Veliidae, Mesoveliidae, Helvidae, Hydrometridae, and Gyrinidae) which share the convenient 2-dimensional visibility and general ease of culture techniques with the Gerridae. This exercise concentrates on Gerris remigis because of its relative ubiquity, its hardiness and amenability to laboratory culture, and because it has been studied more than any other North American species.

Behavioral ecologists are now aware that animals apparently defend fixed or mobile territories only when it makes energetic sense (Krebs and Davies, 1979). Theoretically, this means when the net energy gained by being territorial is greater than the energy gained by not being territorial. In the usual case, one thinks of the gains in terms of mating success. In some cases, however, energy itself -- measured as food intake -- may be the issue. For example, when food is limited but abundant enough above a general maintenance level for the extra energy for defense activity, territoriality may be favored, but as food increases in abundance, a point will eventually be reached where it makes more sense simply to eat and not defend at all, so territoriality ceases (Fig. 25-1).

In the northern United States and Canada, G. remigis overwinters as adults, and begins mating during the warm days of early spring. Breeding males can discriminate between other adult males and females solely by the high frequency ripple signal that males, but not females, produce (Wilcox, 1979). At the time of the spring mating season, territoriality has not yet been observed, though there is reason to believe it may sometimes occur. However, the eggs quickly become nymphs which reach adulthood by mid-June, and during this interval territorial behavior begins to appear within the population. Under conditions of low food supply, territoriality appears among

some nymphs and adults of both sexes. However, among adults many
more males than females are territorial.

This exercise has 2 primary parts. Which activities you
undertake will depend on class interests and season of the year. The
territoriality activities are most successful during the summer and
fall. The sex discrimination activities are most easily done in the
spring, but may be undertaken at any time when G. remigis is mating
in laboratory culture.

METHODS

Subject

Gerris remigis has a distribution from lower Canada down to
Guatemala, is a rather hardy species, and reproduces readily in the
laboratory. It is generally the dominant large (body length, 12-14
mm) water strider on smaller, faster streams, although it is also
found on larger streams and rivers, and sometimes on ponds and lakes.
If you need to identify G. remigis adults and nymphs and/or other
species (keying water striders is usually rather easy), consult the
references mentioned in Aids to the Instructor. Males and females of
G. remigis are easy to distinguish (Fig. 25-2).

Materials Needed

10 or more water striders; laboratory tray set-up (Fig. 25-3);
food source (small, soft arthropods such as flightless
Drosophila); dip net; rocks or bricks; pebbles; air hose; aquarium
bubbler; water circulator mechanism. For sex discrimination
activities only: dissecting microscope; insect pins or straight
pins; Testor's paint or (preferred) tube of black silicone rubber;
soft magnetic stripping; scalpel; tube of contact cement; iridectomy

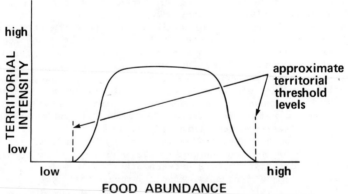

Fig. 25-1. The relationship between territorial intensity and food
levels in a simplified, hypothetical case. The picture, of course,
is usually more complicated in natural situations, due to the
interplay of territoriality with other factors such as mating
strategies.

scissors; 8 ohm point source electromagnet; role of #26 magnet wire (or similar gauge); 8d finishing nail; electric or hand drill; spring-loaded pushbutton switch; 6 mm diam wooden doweling, 20-30 cm long; nonferric forceps (can be made of aluminum foil); small tray; lamp; dark-colored paper (for bottom of tray); function generator.

Procedure

A. Territoriality

The water striders you are studying were marked for individual recognition by your instructor and fed excessively for at least 2 days after their capture. Begin by cleaning the surface of all available food. See to it that no more food can get into the tray. Note the time, and begin to let the striders go hungry.

Working in teams or individually, arrange to observe the striders at the same time once or twice each day. After 1 or 2 days, some (but probably not all) of the striders should begin to establish territories, excluding other striders from rather clearly defined areas. Delineate the open water boundaries by putting lines of pebbles on the bottom, shifting the lines as the strider behavior defines them. Or, for a more quantitative approach, put down a numbered grid on the bottom and score the presence of a strider over particular squares per unit time of observation.

When it occurs, territorial behavior in striders is obvious. There are levels of intensity, however, that can be scored according to the descriptions on the score sheet (Table 25-1). Record occurrence and intensity of territoriality 1 or 2 times per day, until it fades and the territory holders begin to leave their territories and forage for food elsewhere, like the non-territory holders. The striders tend to become a bit listless about this time also, and should be fed before they die. the results of territorial intensity versus days from beginning starvation can now be graphed.

If time permits, there are many manipulations which can be done with this simple system. For example: (1) feed again during territoriality, to note how higher food level affects territoriality;

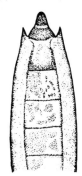 

Fig. 25-2. Dorsal outlines of abdomens of female (left) and male (right) adults of _Gerris remigis_. Note especially the broader abdomen and shorter, more spiky last abdominal segment of the female.

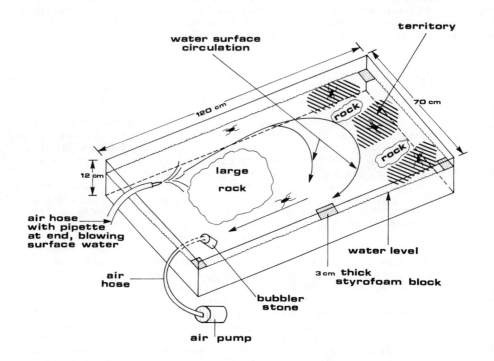

Fig. 25-3. Water strider tray for observing territoriality. The tray can be wooden, waterproofed with plastic sheeting. The bubbler stone is necessary to keep the surface film clean. The water surface should circulate at a rate fast enough to make the striders swim fairly regularly to maintain their station when directly in the fastest current; drift a small cork past a meter stick to determine this (about 4 cm/sec maximum). G. remigis is definitely rheotactic, and will become territorial more readily and behave in general more naturally, when such a current is present.

(2) feed at certain levels per unit time, to determine the maintenance level (see Jamieson and Scudder, 1977); (3) observe territorial takeovers by particular individuals, different sexes, and different nymphal stages; (4) observe recruitment from non-territory holders when territory holders are removed.

B. Sex Discrimination

Males of G. remigis in mating condition can discriminate between other adult males and females solely by the high frequency (HF) ripple signal that other males produce (Fig. 25-4). In the laboratory, these ripple signals can be artificially produced electromagnetically while masking individuals of both sexes to rule out the possibility of visual sex discrimination. If the function generator is battery-operated, this study could also be conducted in the field.

This project is more challenging and exacting in procedure than the territoriality exercise. As preparation for it, study the articles by Wilcox (1979) and Wilcox and Kashinsky (1980) carefully. Then replicate the HF signal playback experiments, altering the techniques described therein in the ways suggested here.

1. Preparing the Apparatus.

The 28 cm diameter coil of #26 magnet wire described in Wilcox (1979) need not be used. Instead, make a circular or square enclosure of about the same size (such as by bending a strip of 3 mm plexiglass into a circle and bolting the overlapping ends together). Make a small "point source" magnet wire coil by putting 12 mm of 8d finishing nail in the chuck of an electric drill, then winding on enough wire (as in Fig. 25-5) to give a resistance of about 8 ohms. Connect the ends of the coil to a function generator, set the generator for 85 Hz, and insert a spring-loaded on-off pushbutton switch.
Attach the nail end to a wooden dowel so the coil can be hand-held vertically over the water.

2. Preparing the Masks.

Both sexes should be blinded for best results (see Wilcox, 1979). The eyes can be painted with Testor's paint. However, it is probably best to make form-fitting rubber masks instead. Smear black liquid silicone rubber (obtainable in small tubes at hardware stores and tropical fish shops) on the head of a dead strider, leaving a narrow "chinstrap" under the head. (Break off the beak of the dead strider first.) Let the rubber cure overnight, then cut the chinstrap across with iridectomy scissors. Peel off the mask. Slip it over the head of a live individual (of the same approximate size) with fine-tipped forceps (see below).

3. Masking the Striders, and Preparing a Mock-Male.

Working under a binocular dissecting microscope, mask 1 or more striders of both sexes. Hold the strider firmly with 1 hand, and slip the mask on snugly, using fine-tipped forceps with the dissecting scope on low power. Make sure the chinstrap is worked into position underneath the beak. For approximate dimensions of the mask, see the issue cover photo for Wilcox (1979).

Then, with a scalpel, cut a tiny piece of magnetic material from some of the soft magnetic stripping, and use contact cement to glue it on top of the femur of 1 foreleg of a G. remigis female (Fig. 25-6). To do this, hold the strider flat on a binocular microscope stage, dab a little glue on top of the femure, then place the magnet chip in place with nonferric forceps. (These can quickly be made by bending a thin strip of aluminum.)

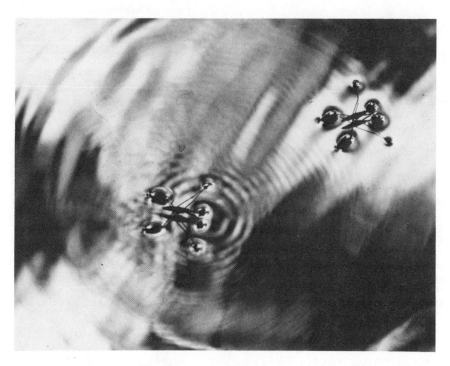

Fig. 25-4. High frequency ripple signals produced by male water striders, <u>Gerris</u> <u>remigis</u>. (From Wilcox, 1980)

As the female swims about in the circle or square, hold the coil directly over her with the nail head pointing down at her, and press the button. By adjusting the power output in the generator and the distance the coil is held over the female (1-2 cm above her is about right), the magnet can be made to cause the foreleg to produce a faint shimmer, about 0.025-0.5 sec long.

The signal intensity and duration are best determined by watching males make HF signals when they are near each other. Each signal is a brief shimmer on the water, made when a male is not actually swimming (Fig. 25-4). The signals can best be seen by placing a sheet of dark paper <u>under</u> the water, then watching the strider when it is sitting in the surface reflection of a lamp placed behind the observation tank.

4. Artificial Manipulations of Water Strider Sex Determination

With a little practice, all the above techniques are quick and surprisingly easy. Males must be in good mating condition, and you must be able to generate signals before males jump on females. As discussed in Wilcox (1979), blinding both sexes tends to increase the "sedateness" of male-female interactions.

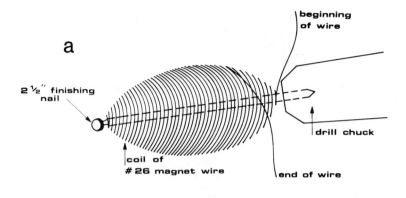

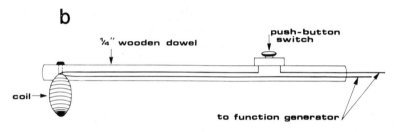

Fig. 25-5. Constructing a device for generating a point-source electromagnetic field oscillated by a function generator. A. Winding the coil to a resistance of 8 ohms. B. Coil mounted on dowel, with push-button switch inserted in one end.

Place males and females together, and carefully follow the signal playback methodology discussed in Wilcox (1979). You will need to read the references and notes carefully in this article, in addition to the text.

SUGGESTIONS FOR FURTHER STUDY

1. Homing Behavior

In the field especially, G. remigis individuals that are holding territories (generally during late June and July in the Northeast, and perhaps on through August or September further south) can "home" back to their territories from a displacement of at least 7.7 m away. (Homing means that an animal returns to a home range or territory after being displaced.) Various distances and directions of displacement could be tried, and the results noted, both with territory holders and with non-territory holders. (Some of the latter are also quite site-specific with essentially mobile territories, day after day.)

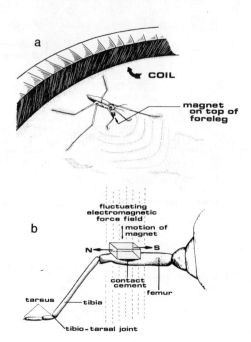

Fig. 25-6. A. Magnet-bearing female of G. remigis producing a
magnet-generated ripple signal inside a large diameter coil. B.
Exact positioning of a tiny magnet on the right fore-femur. (From
Wilcox and Kashinsky, 1980)

2. Drinking Behavior

One fail-safe project that can be done with G. remigis any time
of the year and with striders in almost any internal state is a study
of drinking behavior. Water striders ordinarily have no need for
water conservation and, for an insect, are like leaky sieves. If
left away from water, they will die from dessication overnight.
Water striders drink by assuming a characteristic
head-slightly-down posture while stationary, and extending the beak
straight down to the water. A variety of projects can be done by
noting the drinking rate (number and duration of drinking bouts per
unit time) at a standard relative humidity and temperature, then
comparing drinking rate after various periods (a) without water
available, (b) at different relative humidities and temperatures, and
(c) combinations thereof. Other experiments could explore the
relative influence of air versus water temperature, behavioral
responses to water or land substrate dampness with the striders in
various states of dessication, and so on.

SUMMARY QUESTIONS

1. How is territoriality in Gerris remigis similar to and different
from territoriality in other animals?

2. Why do you suppose territoriality in G. remigis does not appear to occur during the spring mating season (at least in the northern part of its range)?

3. Why is G. remigis territorial in some parts of a stream and not in other parts?

4. Might G. remigis males have other means of discriminating sex besides ripple and contact HF signals? If so, how would you demonstrate the other means of discrimination?

5. What other functions does the HF signal have in the biology of G. remigis?

6. Exactly how are HF signals produced?

SELECTED REFERENCES

Brinkhurst, R.O. 1960. Studies on the functional morphology of Gerris najas Degeer (Hem. Het. Gerridae). Proc. Zool. Soc. Lond. 133: 531-557.

Calabrese, D.M. 1977. The habitats of Gerris F. (Hemiptera: Heteroptera: Gerridae) in Connecticut. Ann. Entomol. Soc. Amer. 70: 977-983.

Carpenter, F.L. and R.E. MacMillen. 1976. Threshold model of feeding territoriality and test with a Hawaiian honey-creeper. Science 194:639-642.

Galbraith, D.F. and C.H. Fernando. 1977. The life history of Gerris remigis (Heteroptera: Gerridae) in a small stream in southern Ontario. Can. Entomol. 109: 221-228.

Gill, F.B. and L.L. Wolf. 1975. Economics of feeding territoriality in the golden-winged Sunbird. Ecology 56: 333-345.

Hungerford, H.C. 1919. Biology of aquatic and semiaquatic Hemiptera. Univ. Kans. Sci. Bull. 11: 1-341.

Jamieson, G.S. and G.G.E. Scudder. 1977. Food consumption in Gerris (Hemiptera). Oecologia 30: 23-41.

Kittle, P.D. 1977. The biology of water striders (Hemiptera: Gerridae) in northwest Arkansas. Amer. Midl. Nat. 97: 400-410.

Matsuda, R. 1960. Morphology, evolution and a classification of the Gerridae (Hemiptera, Heteroptera). Univ. Kans. Sci. Bull. 41: 25-632.

Matthey, W. 1975. Observations sur la reproduction de Gerris remigis Say (Hemiptera, Heteroptera). Mitt. Schweiz. Entomol. Ges. 48: 193-198.

Wilcox, R.S. 1979. Sex discrimination in Gerris remigis: role of a surface wave signal. Science 180: 1325-1327.

Wilcox, R. S. 1980. Ripple communication. Oceanus 23: 61-68.

Wilcox, R.S. and W.M. Kashinsky. 1980. A computerized method of analyzing and playing back vibratory animal signals. Behav. Res. Meth. Instr. 12: 361-363.

TABLE 25-1. Sample data collection sheet for studying territorial behavior in <u>Gerris</u> <u>remigis</u>, intended to be used mainly as a checklist, with additional notes taken as seems sensible.

Developmental Stage:
(___instar nymph, adult)

Date:
Time: to
Air Temperature:
Water Temperature:

      (Draw diagrammatic strider,
      then indicate color code
      of identifying paint spots)

Individual Strider: (Strider A, B, etc.)
Territory Number:

		BEHAVIORAL CATEGORIES:*			
Encounter Number	No Reac.	Swim At	Lunge At	Jump On	NOTES
1					
2					
3					
etc.					

---

*Key to Behavioral Categories on Checklist

Encounter: 2 individuals, 1 being territorial, encounter each other.
No Reaction: Territory holder shows no reaction to presence of intruder.
Swim At: Territory holder swims, usually rapidly, at intruder, usually causing intruder to retreat.
Lunge At: Territory holder swims at an intruder, then rapidly lunges its body back and forth at the intruder, usually causing intruder to retreat.
Jump On: Territory holder swims at (up to) intruder and jumps up onto intruder's back, usually causing intruder to retreat. Territory holder may or more commonly may not lunge at intruder first.

(NOTE: These behavioral categories indicate increasing levels of aggression by the territory holder, although "lunge at" and "jump on" may represent more or less comparable levels of aggression.)

BROOD CARE, NESTING,
AND SOCIAL LIFE

26. Foundress Interactions in Paper Wasps (*Polistes*)

Robert W. Matthews
University of Georgia

The evolution of non-reproducing individuals in social insects has been problematic to biologists for some time. In the last 20 years, a number of theoretical explanations have been argued, some stressing genetic, others proximate environmental factors (for a review, see Starr, 1979). Because both types of arguments generated biologically testable predictions, they have spurred a great deal of interest in basic social insect biology.

Among the social Hymenoptera, paper wasps are considered to be a key group. Their spring nest-founding ritual has been the the subject of special interest, for the annual nests are begun either singly (haplometrosis) or jointly (pleometrosis) by a clique of fertile females called foundresses. In many pleometrotic colonies of polistine wasps, these females have been shown to contend for the queen role, eating one another's eggs (oophagy) and replacing them with their own. One contender ultimately becomes the sole egglayer for the colony, while her subordinates disappear or become function-ally sterile auxilliary helpers. A very readable account of polistine biology may be found in Evans and Eberhard (1970).

In this exercise, we undertake a study of <u>Polistes</u> behavior in the field during a spring afternoon, observing foundress activities in detail. At the end of the study, the marked foundresses will be collected for ovarian dissection in order to examine the relationship between their behavior and reproductive condition.

METHODS

<u>Subject</u>

Paper wasps belong to the family Vespidae, which also includes yellowjackets and hornets, all of which are social. Colonies are composed of many sterile workers, an egglaying queen, and at certain seasons, numerous males or drones.

Although all vespids are technically "paper wasps" because they build nests using pulp from macerated plant fibers, the wasps most commonly called by this name belong to the genus <u>Polistes</u>, which build easily observed nests having a single comb of hexagonal cells without an envelope -- the small wasp nest commonly found in barns and garages and beneath the eaves of houses. Because their activities take place in plain view, and because colony size is relatively small, these social insects have been popular research subjects (see Selected References).

Materials Needed

Notebooks; clipboards; stopwatches or wrist watches with second hand; jars with lids; dissecting microscopes; wax-bottom dissecting dishes; fine-pointed scissors or single edged razor blades; insect pins; various colors of fast drying paints; brushes. Optional: portable cassette tape recorders, ladder.

Procedure

For the field portion of this exercise, individuals or pairs of students will be assigned to each nest in a place suitable for detailed observations. Most nests will likely be located on old buildings, such as sheds and barns. All foundresses on the nests to be observed will have been marked prior to class. Take care that your presence does not interfere with colony activities. If possible, at least 1 nest having only a single foundress should be included in the observation nests in order to obtain comparative data.

NOTE: Foundress females are not aggressive at this time of year, and normally can be observed closely with little danger. Nevertheless, they can sting and will attack if provoked. Persons known to be allergic to insect stings or bee venom should not participate in the field portion of this exercise.

A. Nest Information

Background: Nests should be approximately 2 to 4 weeks old, and contain several cells with eggs and some with larvae. Depending on species and temperatures, eggs require about 2 weeks to hatch; the larval stage averages 15 days before the appearance of a gleaming white silken cap over the cell signals pupation. After about 22 days inside the capped cell, a fully pigmented adult will emerge. All adults present in spring are fertile overwintered females, each potentially able to become a full queen.

For the nest to which you have been assigned, record at least the following information: nest location (height, distance from corners, distance from other nests, etc.); number of cells, eggs, and young larvae (can be deferred until nest is collected); number of adults (foundresses); individual recognition marker of each foundress present.

B. Time And Motion Study

For at least 1 hour, keep a continuous, timed record of activities observed for each individual on the nest (Table 26-1). To facilitate record keeping, you might wish to arbitrarily assign each wasp a number coded to its color mark. Be especially alert for any behaviors that might be associated with queen dominance, such as tail-wagging (rapid bursts of side to side abdominal movement; see Gamboa and Dew, 1981), postural attitudes, or egglaying. Some behaviors, such as cell inspections, may need to be lumped into bouts, during which several cells are rapidly inspected, one after another. For purposes of this exercise, such a series of inspection visits could be considered as one unit of behavior. Grooming of antennae, mouthparts, and legs could similarly be lumped and simply called "grooming". The important distinctions are the transitions between categories of behavior, as suggested by the following list of behaviors to be recorded:

1. Foraging trips by each wasp (duration, and if possible outcome)
2. Solicitation and food exchange among nestmates
3. Cell inspection/cell antennation
4. Biting (and response) among nestmates ("aggression")
5. Oviposition
6. Cell initiation or enlargement
7. Brood-feeding
8. Wing-flipping (rapid wing flutter)
9. Grooming
10. Resting
11. Tail-wagging
12. Gaster-smearing*
13. Rapid, apparently undirected movement

At the end of the class period, make an informal guess ranking the members of your foundress clique from most dominant (queen) to least dominant, based on the observed behaviors. Then collect all wasps from your nest. Foragers can be netted as they return. Individuals on the nest may be captured as the nest is collected.

To capture the nest, swiftly slip a wide mouth glass jar over the nest, pressing the jar tightly against the ceiling or overhang. Slide it sideways to break the nest pedicel, then slip an index card or screw lid between the jar and the overhang, capping the jar. Nest and adults should be killed by freezing, and stored in a freezer until analysis.

C. Ovarian Development And Individual Size Determination

In the laboratory, obtain a dissecting microscope, wax-bottom dissecting dish, fine pointed scissors, and insect pins. Lay 1 wasp on its back in the dish in enough water to submerge its abdomen, and pin it in place through the thorax. Gently cut and/or remove the ventral sclerites from the abdomen. Spread and pin the dorsal sclerites to each side to expose the reproductive system.

For descriptive and comparative purposes, a 4-class scale of ovarian development (OD) has been found useful (Fig. 26-1). Dissect and classify each foundress from your nest according to this system,

---

*Rubbing venter of abdomen on nest or petiole; a sternal gland is source of repellent secretions (See Hermann and Dirks, 1974).

TABLE 26-1. Sample data sheet for a time and motion study of
Polistes behavior.

Species: Observer's name:

Nest location:

Nest composition: No. of cells ____   No. of eggs ____
                   No. of larvae ___   No. of pupae ___
                   No. of foundresses _____

Temperature (°C):   Sky condition:

Foundress No. and distin- guishing marks	Time		Behaviors (continuous record)
	Begin	End	

wasp 1 -

wasp 2 -

  etc.

remembering to keep track of individual markings. The queen is considered to be the female whose ovaries are most developed. On the
basis of ovarian development, attempt to rank the females from your
nest. (To improve data resolution, you might wish to measure the
length of each female's longest oocyte.)
     As an indicator of comparative size, wing length has been found
to be a convenient measure highly correlated with a number of other
measures such as head width and body length. Prior to discarding the
wasps, measure and record forewing length for each one.

D. Data Analysis

     A number of interesting descriptive and inferential statistical
analyses are possible. Those chosen should be run independently by
each student or team. Share team results so everyone has access to
all the descriptive data. Some suggestions follow:

1. Descriptive Statistics

     a. Calculate mean, standard error and range for the total class
sample for the following nest parameters: number of foundresses,
cells, eggs, larvae and capped brood.

   b.   Tabulate for each foundress both the frequency (number of
occurrences) and duration for each behavioral act.   Break these data
down according to foundress rank and compare mean, standard error and
range of the total queen sample with the same data for those females
determined to be subordinate based on ovarian development.
   c.   Similarly compare the mean OD indices for all queens and subordinate foundresses.
   d.   Try to determine dominance hierarchy by analyzing your data
in the manner described in exercise 18 (see Table 18-1, p. 123).

2.   Inferential Statistics

   a.   Use the Mann-Whitney U test (see Appendix 1.4) or Student's
t-test to compare the mean wing lengths of queens and other foundresses.   Are the 2 significantly different?
   b.   Calculate the correlation coefficient of wing length and
length of longest oocyte (see Appendix 1.6, Kendall's tau).   How good
is the correlation between these 2 parameters?
   c.   If the data are available, compare mean time spent away from
the nest for queens on multifoundress nests and lone-nesting females.
Similarly compare the same for queens and subordinate cofoundresses.
(Note: you could make these comparisons using either the frequency or
duration data.)   If your data indicate the outcome of particular
foraging trips it might also be possible to learn whether there are
differences in the number and/or duration of trips for pulp versus
flesh or fluid.
   d.   If you made careful notes, it may be possible to compare

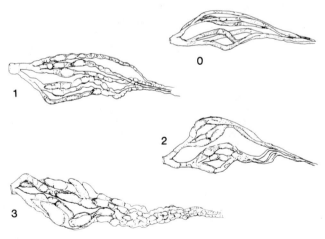

Fig. 26-1.   Four stages of ovarian development in Polistes females
(after Litte, 1977).   0 -- undeveloped;   1 -- one or both ovaries
slightly enlarged;   2 -- one or both ovaries considerably enlarged,
with several small or medium-sized oocytes;   3 -- both ovaries
enlarged, with several mature oocytes (the queen condition). (Drawing
by David Mallow).

data on the nature of solicitation. One common type is for foundresses to beg from foragers returning with nectar or flesh. Another
common maneuver is begging directly from larvae (often in turn being
obliged to pass this fluid on to others). Some researchers have
noted that queens appear to channel a considerable amount of the
available food to and through themselves. Were the number of
solicitations you observed involving the queen significantly
different from the number involving other foundresses on a
particular nest?

SUGGESTIONS FOR FURTHER STUDY

The exercise focused on data obtainable in a single afternoon.
If the opportunity exists for longer term studies, other interesting
questions may be asked.

1. Nest Productivity

Collect data on size (number of cells) and foundress number for
nests at regular intervals (weekly if possible) over the spring until
the time of first adult emergence. Graphically determine whether
there appears to be any relationship between nest size and number of
foundresses. Calculate productivity on a per-foundress basis. How
might the results be explained?

2. Survivorship

Mark all foundresses on several nests, and follow the nests
through until first worker emergence. Compare the mean probability
of survival for single-foundress versus multi-foundress nests. The
Fisher exact probability test (Appendix 1-3) may be used to analyze
this. How much turnover in foundress clique makeup did you detect?

3. Alternative Nesting Strategies

In many parts of the country haplometrotic species nest side by
side in the same habitat with pleometrotic species. If such is the
case in your area, comparative data on all aspects discussed above
provides an opportunity for evolutionary speculations.

4. Nest Repair

Simulate an attack on various nests by a bird by carefully
cutting out a triangular "peck-sized" area from an active nest. (Work
carefully and quickly after first chasing the wasps off the nest.)
Observe the wasps' response to this disturbance over the next several
days and record what you see. If the wasps repair the break is it
done in such a way as to restore the nest symmetry?

5. Nest Defense Against Vertebrates

Colony defense is a key function of social insect behavior and
it has been suggested that one benefit to foundress groups is

improved defense (see Gibo, 1978 and Gamboa, 1978). Because Polistes show a number of distinct and easily evoked threat behaviors in response to disturbance, it is possible to also rank the members of a foundress group according to the strength of their response to provocation (C. K. Starr, personal communication).

One especially useful behavioral response to mild provocation is the leg wave, recognizable when a wasp elevates her body slightly and raises her front legs off from the nest and simultaneously waves them from side to side about 10 times a sec. As a standardized provocation, slowly approach an undisturbed nest waving a net or net handle slowly back and forth. As the proximity of the net handle to the nest increases, keep track of the order in which the marked individual wasps initiate leg waving. Replicate 2 or 3 times until the order is known. (Note: because of the potential for actually provoking an attack, the provoker should wear protective clothing and a bee veil while performing this experiment.) At the conclusion of the experiment collect all the foundresses and determine their ovarian development. Compare the rankings obtained by the 2 methods.

As one increases the intensity of provocation individual wasps will likely display a hierarchy of defensive responses. The defensive repertoire and the sequence in which the various acts occur may differ between species, providing an opportunity for a potentially rewarding comparative study.

SUMMARY QUESTIONS

1. How well did your ranking of the nest foundresses based upon behavioral attributes compare to the ranking obtained as a result of ovarian development? How reliable is behavior alone as an indicator of reproductive status?

2. Was there a clear-cut dominance hierarchy among the foundresses in your colony? If so, what benefits of such an arrangement can you postulate?

3. From the foundresses' standpoint, generate a list of possible advantages and drawbacks resulting from participation in foundress cliques.

4. Consider the locations of the nests your class studied and attempt to identify the attributes of an "ideal" potential nest site. Does there appear to be any relationship between nest location and the number of foundresses or nest size?

5. You have probably observed nests built on man-made structures. Where would you look for nests of your wasp in nature?

SELECTED REFERENCES

Eberhard, M. J. W. 1969. The social biology of polistine wasps. Misc. Publ. Mus. Zool., Univ. Mich., no. 140. 101 pp.
Evans, H. E. and M. J. W. Eberhard. 1970. The Wasps. Univ. Mich.

Press, Ann Arbor. See espec. pp. 126-154.

Gamboa, G. J. 1978. Intraspecific defense: Advantage of social cooperation among paper wasp foundresses. Science 199: 1463-1465.

Gamboa, G. J. and H. E. Dew. 1981. Intracolonial communication by body oscillations in the paper wasp, Polistes metricus. Insectes Sociaux 28:13-26.

Gibo, D. L. 1974. A laboratory study of the selective advantage of foundress associations in Polistes fuscatus. Can. Entomol. 106: 101-106.

Gibo, D. L. 1978. The selective advantage of foundress associations in Polistes fuscatus. A field study of the effects of predation on productivity. Can. Entomol. 110:519-540.

Hermann, H. R. and T. F. Dirks. 1974. Sternal glands in polistine wasps: morphology and associated behavior. J. Georgia Entomol. Soc. 9: 1-8.

Hermann, H. R. and T. F. Dirks. 1975. Biology of Polistes annularis. I. Spring behavior. Psyche 82: 97-108.

Klahn, J. E. 1979. Philopatric and nonphilopatric foundress associations in a social wasp Polistes fuscatus. Behav. Ecol. Sociobiol. 5:417-424.

Litte, M. 1977. Behavioral ecology of the social wasp, Mischocyttarus mexicanus. Behav. Ecol. Sociobiol. 2:229-246.

Nelson, J. M. 1968. Parasites and symbionts of nests of Polistes wasps. Ann. Entomol. Soc. Amer. 61: 1528-1539.

Noonan, K. M. 1981. Individual strategies of inclusive fitness maximizing in Polistes fuscatus foundresses. In R. D. Alexander and D. W. Tinkle (eds.) Natural Selection and Social Behavior: Recent Research and New Theory. Chiron Press, New York. pp. 18-44.

Pardi, L. 1948. Dominance order in polistine wasps. Physiol. Zool. 21:1-13.

Starr, C. K. 1979. Origin and evolution of insect sociality: A review of modern theory. In H. R. Hermann (ed.) Social Insects, vol. 1, Academic Press, New York, pp. 35-79.

Strassmann, J. E. 1981a. Wasp reproduction and kin selection: Reproductive competition and dominance hierarchies among Polistes annularis foundresses. Fla. Entomol. 64: 74-88.

Strassmann, J. E. 1981b. Kin selection and satellite nests in Polistes exclamans. In R. D. Alexander and D. W. Tinkle (eds.) Natural Selection and Social Behavior. Recent Research and New Theory. Chiron Press, New York. pp. 45-58.

West, M. J. 1967. Foundress association in polistine wasps: Dominance hierarchies and the evolution of social behavior. Science 157:1584-1585.

27. Sound Production and Communication in Subsocial Beetles

William H. Gotwald, Jr.
*Utica College
of Syracuse University*

In social insects, the behavior patterns of individuals are woven together to produce the apparently cooperative mass actions that characterize social behavior. The weaver's shuttle that crafts this fabric of sociality is communication (Wilson, 1971). For most social insects, the primary communication mode is by way of chemical messengers called pheromones. A pheromone is released into the environment from an insect's body and produces predictable responses among the releasing insect's nestmates. Although visual signals and airborne sound appear to be relatively unimportant in the communication systems of social insects, substrate vibrations may be consequential for certain species. Some termites, for instance, bang their heads on the substrate, producing vibrations that signal alarm (Stuart, 1969).

Since communication is essential to the cohesive interactions that typify social insects, it is instructive to investigate communication modes and systems in subsocial species, where the adults care for their immature offspring for at least some period of time, and in primitively social species. Such studies of presocial species help expand our understanding of the evolutionary origin and nature of sociality.

Many species of the beetle family Passalidae are subsocial. Commonly known as bess beetles, these robust insects are most numerous in tropical environments. Only 3 species are recorded from the United States (Arnett, 1968). The most common of these is Popilius disjunctus, a glossy black species that is distributed throughout the eastern half of the United States. This species spends most of its life cycle tunneling in rotting hardwood logs and stumps. The adults are gregarious, and live in groups or "colonies" that eventually include the larvae as well. The entire life cycle from egg to adult takes 2½ to 3 months to complete. Only 1 generation of new adults is produced each year. Endowed with insatiable appetites, the larvae feed on rotting wood that has been chewed by the adults and processed into a pulpy "frass". Although the larger, older larvae have mandibles capable of biting into rotting wood, in order to be fully nourished they apparently require the adult-prepared frass (Pearse et al., 1936; Gray, 1946).

Thus bess beetles are communal, and the adults provide, or at least make accessible, food for the larvae. They are social to this extent. They also communicate. Both adults and larvae possess organs for stridulation -- the process of producing sound by rubbing 2 body surfaces together. In the larvae, the stridulatory mechanism consists of a modified metathoracic or hind leg that is rapidly rubbed over a "striated" area on the mesothoracic or middle coxa (Gray, 1946; Reyes-Castillo and Jarman, 1980). Although the sound produced by this mechanism is weak, it is still audible to the human ear. The adult stridulatory structures include specialized sections of the wing-folds (where the wings bend back upon themselves, and thus fit snugly beneath the elytra or wing covers) and 2 ovate areas on the dorsal surface of the fifth abdominal segment. When the adult beetle stridulates, the abdomen is elevated, pressing the wings against the inside surface of the elytra. This movement forces the wing-folds to rub across the surface of the ovate areas (Babb, 1901). The sound produced is distinct and loud.

What function can this sound play? Most significantly, it was noted at the beginning of this century that these beetles seemed to stridulate only when disturbed (Babb, 1901). Investigators have speculated that these sounds may serve as a danger signal to other members of the colony, as a warning to intruders from other colonies, and/or as a means of "keeping the colony together" (Gray, 1946). Recordings of these stridulatory sounds have revealed that, in addition to the sound produced when adults are disturbed, there is a distinct sound created by adult aggressors when fighting other adults (Alexander et al., 1963). But little is really known about the behavior of bess beetles -- so little, in fact, that other functions for these sounds, such as in sexual behavior, cannot be ruled out.

METHODS

Subject

Popilius (now correctly called Odontotaenius) disjunctus, an eastern species of the family Passalidae, is a shining black beetle, 32-36 mm long, with longitudinal grooves in its elytra and a characteristic horn on its head (Fig. 27-1). Colonies are fairly common in galleries in decaying oak logs.

Materials Needed

Live adult bess beetles; dissection instruments; microscope slides and cover glasses; small dissection dishes with paraffin bottoms; pins for immobilizing specimens; carbon dioxide source and anesthetizing chamber (a bottle provided with a 2-hole stopper and tubing to permit a constant flow of CO_2 through the bottle); insect Ringer's solution (for supplementary observations); compound and dissecting microscopes.

Procedure

Select live adult specimens of P. disjunctus from the laboratory population. To hear the sound produced by bess beetles,

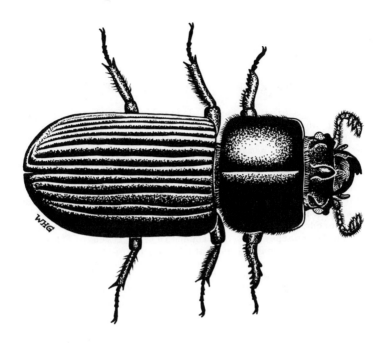

Fig. 27-1. <u>Popilius</u> <u>disjunctus</u> may be called by a variety of common names -- bess bugs, betsy beetles, patent-leather beetles, and horned passalus beetles. Adults appear to have a rudimentary social organization including aggressive interactions which may often be accompanied by audible stridulatory noises.

at least initially, pick up a beetle (they do not bite) and hold it close to your ear. Just grasping a specimen should constitute enough disturbance to stimulate stridulation.

Identify and observe the stimuli that elicit stridulation in these beetles. Can you provoke 2 adults into fighting? If so, can you detect differences between the aggressive stridulatory response and the common disturbance response?

Begin locating the stridulatory structures by observing abdominal movements during stridulation. Devise and perform a series of experiments that will identify the stridulatory structures. (Eventually you may want to cut away portions of the elytra and wings when narrowing down the location of the stridulatory structures.) Describe and illustrate the surface morphology of the ovate areas. How is sound produced?

When you have located the ovate areas, anesthetize the beetle with carbon dioxide and cut these areas from the abdominal sclerite. Make a simple "wet mount" of these excised areas on a microscope slide so that you can examine the fine structure responsible, in part, for the stridulatory sounds. Describe and illustrate.

Formulate a hypothetical laboratory and/or field study that might elucidate the nature and function of stridulation in this species.

SUGGESTIONS FOR FURTHER STUDY

Adults of P. disjunctus are literally "crawling" with a variety of commensal and parasitic animals -- they carry about a veritable zoo of such organisms. Eighteen guest species are known, and include 14 mites, 2 nematodes, a protozoan, and a fly (Pearse et al., 1936; Hunter and Mullin, 1964; Harrel and Mathis, 1964; Harrel, 1967).

A majority of the mite species are distributed in the concavities beneath the head and in crevices around the legs. Two species are found on the wings and inside surfaces of the elytra. Of the nematode species, 1 inhabits the hemocoel, often in large numbers, while another can be found in the lumen of the alimentary canal. Gregarina passalicornuti, a parasitic protozoan, can sometimes be found in the midgut (or ventriculus) of the digestive tract, attached to the midgut wall or floating about in the lumen (Pearse et al., 1936). The parasitic fly, Zelia vertebrata, infests the beetle larvae and can be found as maggots inside their hosts. This fly probably accounts for the high mortality rate among larvae (Gray, 1946).

If time permits, you may want to pursue 1 or more of the following observations and suggestions:

1. Examine an adult specimen under the dissecting microscope for the mites of the head and leg regions. Specimens of mites can be removed from the beetle with a dissecting needle. Keep a record of where each mite was located on the beetle, and mount the mites on microscope slides. Examine these mites with the compound microscope. (Mite morphology is fascinating and at times bizarre.) How many species of mites have you found? Is there any species specificity in their patterns of distribution on the beetle?

2. Remove the elytra and wings, and examine these structures for the 2 species of mites that commonly reside there. On a specimen from which the elytra and wings have been removed, note the pulsating dorsal vessel visible through the transparent dorsal wall of the animal. This vessel pumps the "blood" (or hemolymph) into the head region, from which it percolates back into the hemocoel of the thorax and abdomen. Under the dissecting microscope, observe this "wingless" adult for the nematodes it harbors. These usually can be seen through the transparent dorsum, wiggling about in the hemocoel.

3. Under insect Ringer's solution, dissect open the hemocoel of an anesthetized beetle. Isolate and count or estimate the number of nematodes (Chondronema passali) in the hemocoel. (Pearse et al., 1936, counted 4,260 nematodes in a single beetle!) Remove the beetle's alimentary canal. Open the canal under the Ringer's solution and examine the lumen for the nematode, Histrignathus.

SUMMARY QUESTIONS

1. What might be the function of stridulating when disturbed? How could such communication benefit the stridulator or its colony-mates?

2. What kinds of messages could be "packaged" in the stridulation of an adult passalid aggressor?

3. Do you suppose that the functions of stridulation in the larvae are the same as those of adult stridulation? Speculate about the possible functions of larval stridulation.

4. List possible communicative functions for adult stridulation in addition to those involved in disturbance and aggression.

5. When some researchers concluded that stridulation in Popilius disjunctus was a way of "keeping the colony together", what do you think they meant?

6. Can you devise an evolutionary scenario that explains the origin of stridulation in passalid beetles?

7. What are the behavioral differences that separate subsocial and truly social (eusocial) insects?

8. Why would beetles become subsocial? That is, what are the adaptive advantages to the beetle of evolving a subsocial behavioral repertoire?

9. Why do you think most social insects have evolved reliance on chemical communication rather than on alternative possibilities such as airborne sound?

SELECTED REFERENCES

Alexander, R. D., T. E. Moore, and R. E. Woodruff. 1963. The evolutionary differentiation in stridulatory signals in beetles (Insecta: Coleoptera). Anim. Behav. 11: 111-115.
Arnett, R.H. 1968. The Beetles of the United States. (A Manual for Identification). Amer. Entomol. Inst., Ann Arbor, Mich. 1112 pp.
Babb, G.F. 1901. On the stridulation of Passalus cornutus Fabr. Entomol. News 12: 279-281.
Gray, I.E. 1946. Observations on the life history of the horned passalus. Amer. Midl. Nat. 35: 728-746.
Harrel, R.C. 1967. Popilius disjunctus (Illiger) (Coleoptera: Passalidae) as a laboratory animal. Turtox News 45: 270-271.
Harrel, R.C. and B.J. Mathis. 1964. Mites associated with Popilius disjunctus (Illiger) (Coleoptera: Passalidae) in McCurtain County, Oklahoma. Proc. Okla. Acad. Sci. 45: 66-68.
Hunter, P.E. and K. Mullin. 1964. Mites associated with the passalus beetle. I. Life stages and seasonal abundance of

Cosmolaelaps passali n. sp. (Acarina: Laelaptidae). Acarologia 6: 247-256.

Pearse, A.S., M.T. Patterson, J.S. Rankin, and G.W. Wharton. 1936. The ecology of Passalus cornutus Fabricius, a beetle, which lives in rotting logs. Ecol. Monogr. 6: 455-490.

Reyes-Castillo, P. and M. Jarman. 1980. Some notes on larval stridulation in Neotropical Passalidae (Coleoptera: Lamellicornia). Coleop. Bull. 34: 263-270.

Stuart, A.M. 1969. Social behavior and communication. Biology of Termites. In K. Krishna and F.M. Weesner (eds.), vol. 1, Academic Press, New York, pp. 193-232.

Wilson, E.O. 1971. The Insect Societies. Belknap Press of Harvard Univ. Press, Cambridge, Mass.

Part I

INTEGRATION AND ORGANIZATION OF BEHAVIOR

28. Comparative Anatomy of the Insect Neuroendocrine System

John G. Stoffolano, Jr.,
and Chih-Ming Yin
University of Massachusetts

How an insect responds to external and internal stimuli depends not only on the nature of the stimulus, but also on the insect's physiological state. Under natural conditions, behavior is a complex mixture both of stereotyped responses to stimuli and of learning.

Two distinct but closely interconnected communication systems, the nervous system and the endocrine system, mediate insect behavior (Table 28-1). The nervous system is electrochemical, its message being provided by nervous impulses traveling along the nerve cells (neurons). These follow specific pathways which depend on connections between the neurons (synapses). Conduction of the nervous impulses across the synapses can take place in 2 ways: (1) the release of a chemical can depolarize the postsynaptic area, producing electrotonic currents that evoke nervous impulses in the adjacent portion of the neurons (chemical synapses); (2) some neurons are just electrically coupled, and depolarization of the postsynaptic area results by an electrical event (electrical synapses).

Whereas the nervous system is mainly involved in short term communication, the endocrine system functions to coordinate long term events in the life of the insect. The endocrine system is chemical. It uses specific substances (hormones) produced and released by certain cells, then transported throughout the body until they reach certain other uniquely sensitive cells and organs (called target sites) which respond to them.

Specialized cells called neurosecretory cells (Fig. 28-1), are found throughout the insect's central nervous system, and are identified either by using a selective stain and/or by using transmission electron miscroscopy. These cells are unique to both the nervous and endocrine systems since they generate spontaneous electrical activity and also secrete a chemical that usually travels via the hemolymph to target sites. The 2 systems are closely connected, both structurally and functionally, and interact in a cooperative fashion to determine the behavioral and physiological responsiveness of an insect to external and internal stimuli during its life cycle.

In this exercise, you will have the opportunity to identify and compare the neuroendocrine system of a primitive type of insect, the

TABLE 28-1. The neuroendocrine system -- the internal programming
and integration system mediating insect behavior and development.

Name	Usual Location	Role(s) (often incompletely known)
I. CENTRAL NERVOUS SYSTEM (CNS)		
a. brain (supra-esophageal ganglion)	main ganglionic mass (located above esophagus and behind eyes) composed of a no. of fused ganglia and divided into 3 main regions: protocerebrum, deutocerebrum and tritocerebrum.	coordinates many activities involving the entire body; contains neurosecretory cells that extend to corpora cardiaca and corpora allata; forms bridge with, and is main association center between, sensory input from antennae, eyes, other sensory receptors, and the motor output.
b. sub-esophageal ganglion	lower part of head below the esophagus	supplies nerves to mouthparts, neck, and salivary glands; controls release of some competing behavior modes through neural inhibition.
c. ventral nerve cord	paired nerve, segmental ganglia, and nerves to peripheral sense organs and muscle systems of thorax and abdomen.	coordinates information to and from peripheral sense organs and muscle systems of thorax and abdomen; neurosecretory cells in the segmental ganglia form a bridge between nervous and endocrine systems.
d. stomadeal nervous system	2 median ganglia (frontal and hypocerebral) connected to the brain.	regulates foregut and midgut activities (i.e. swallowing); also called "visceral", "stomatogastric" or "sympathetic" nervous system.
II. PERIPHERAL NERVOUS SYSTEM		
a. peripheral nerves	sensory and motor neurons between CNS and body surface.	communicate with integument and muscles.
b. sensory organs and receptors	eyes, antennae, and various sensilla, etc.	gather external information while filtering out nonappropriate stimuli.

TABLE 28-1, cont.

Name	Usual Location	Role(s) (often incompletely known)

III. ENDOCRINE SYSTEM

a. prothoracic, thoracic, or ecdysial glands	in the thorax or at the back of the head.	produce ecdysone which initiates growth and development and causes molting; normally glands break down soon after final molt to adult.
b. corpora cardiaca	pair of glands closely associated with, and often forming part of, the aorta wall.	store and release neurosecretions from the neurosecretory cells of the brain, and also produce neurosecretions that influence behavior and physiological processes.
c. corpora allata	small glands immediately behind the brain on either side of the esophagus.	produce juvenile hormone which influences direction of molt and promotes yolk deposition in eggs; c.a. secretions have been linked with various reproductive and metabolic processes; have nervous connections with brain, and nerve impulses affect their activity.
d. ring gland (Weismann's ring)	found in cyclorrhaphous Diptera; surrounds aorta just above the brain.	formed by fusion of corpora allata, corpora cardiaca, and prothoracic glands; connected to brain by pair of nerves. Presumably plays same roles as individual non-fused glands.
e. neurosecretory cells	found both in the brain and in the ganglia of thorax and abdomen.	produce neurosecretions which act directly upon a variety of effector organs; as intermediates they produce neurosecretions that stimulate other endocrine glands to secrete hormones.

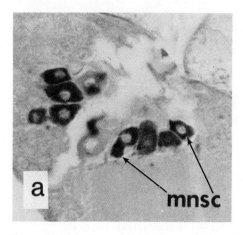 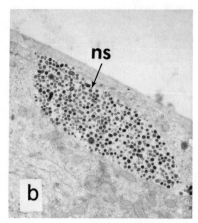

Fig. 28-1. Neurosecretory cells. A. Light micrograph of the brain of the blow fly Phormia regina, showing the median neurosecretory cells (mnsc) that are differentially stained with paraldehyde fuchsin. B. Electron micrograph showing the electron dense, neurosecretory particles (ns) in the axon of an insect.

cockroach, and a more advanced type, the fly. You will observe the neuroendocrine system of 2 life stages of the fly, the larva and the adult, to note differences which occur during metamorphosis. In the process, you will gain skills in the technique of vitally staining the nervous tissue of insects, and will learn the location of the endocrine glands in the cockroach and fly so that, if necessary, you could perform gland extirpations and implantation. The latter 2 techniques are important to behaviorists who are trying to establish the role of insect hormones on behavior (e.g. Truman and Riddiford, 1974).

METHODS

Subject

Larvae and pupae of the blow fly (Phormia regina) or flesh fly (Sarcophaga bullata) may be obtained from biological supply houses, bait stores, and pet shops. The house fly, Musca domestica, can also be used if available. Biological supply companies also stock the American cockroach (Periplaneta americana). If you wish to culture these organisms yourself, consult Galtsoff et al., 1959.

Materials Needed

Fly larvae and adults; cockroach nymphs and adults; dissecting tools, including single-edged razor blades, forceps, insect pins, scissors, and wax-bottom dishes; dental wax; disposable pipette; gas burner; ice cubes; petroleum jelly; paper towels; microscope slides and cover slips; dissecting microscope; phase contrast microscope; methylene blue solution; saturated ammonium molybdate solution; ethyl alcohol in graded concentrations (see part A); 0.75% NaCL solution; xylene; permanent slide mounting medium (optional).

Procedure

In this exercise, you will be examining various parts of the insect neuroendocrine system. As necessary, refer to Table 28-1 to keep the "total picture" in your mind as you work.

For several parts of this exercise, you need to be able to use the technique of vitally staining with methylene blue, outlined below. This technique has been widely used by neuroanatomists, and is extremely important for mapping the insect's nervous system.

1. With a gas burner, heat a glass pipette, and pull the tip to a fine point.

2. Immobilize a fly by chilling it over an ice cube.

3. Pin the fly, dorsal side up, on dental wax placed over the ice cube.

4. With the fine-tipped pipette fitted with a compression bulb, inject the methylene blue solution through the medial-dorsal region of the prothorax.

5. Leave the insect on ice for 10-20 min. Inject it a second time, and wait 10 min. more.

Another technique that has proven to be indispensable in mapping neural circuitry inside ganglia is intracellular staining. If you are interested in this technique, refer to the books by Kater and Nicholson (1973), Miller (1979), and Strausfeld and Miller (1979).

A.   Observing the Peripheral Nervous System

Following the procedure outlined above, vitally stain an adult fly. Then gently pull out the mouthparts, and place them in a vial containing saturated ammonium molybdate solution. This solution will form a fine granular precipitate with methylene blue in the tissue. Leave overnight. Dehydrate the mouthparts in graded series of ethyl alcohol (30%, 70%, 95%, and 2 changes of absolute). Clear in xylene. If you want a permanent slide, mount in some permanent mounting medium.

Your instructor may have chosen to treat this step as a demonstration. If so, see the demonstration slides of peripheral nerve cells and nerves stained with methylene blue stain.

B.   Observing the Fly's Central Nervous System and Endocrine System

Begin by vitally staining an adult fly, following the procedure outlined above, if you have not already done so for Part A.

With the fly pinned dorsal side up on dental wax over an ice cube, laterally cut away the abdominal tergites. Add drops of saline solution (0.75% NaCL) to cover the specimen. (Dissections are easier to perform if specimens are covered in saline; this causes the tissues to float so that they do not adhere to one another.) Carefully remove the digestive tract, trachea, fat body, etc., from the abdominal cavity until the <u>ventral nerve cord</u> (a large nerve trunk with a few branches on the side) is exposed. Refer to Fig. 28-2 as necessary.

Cut dorsal-medially through the thoracic tergites. This takes time, since you have to cut through and remove the extensive thoracic flight muscles. Take special care near the neck region until the fused <u>thoracico-abdominal ganglion</u> is exposed.

Carefully cut open the cranium through the occipital foramen, vertex, frons, and clypeus. Break away and remove the cranium to expose the <u>brain</u>.

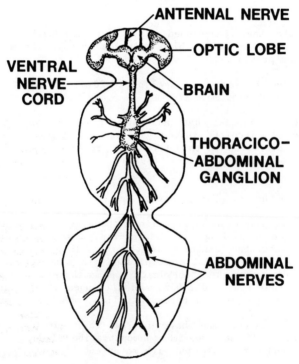

Fig. 28-2. Drawing of the central nervous system of an adult fly with the parts labelled.

Remove saline, and add a few drops of methylene blue solution on the specimen. The nerve branches from the central nervous system should pick up the most stain. After 1-2 min., wash away excess stain with 0.75% NaCL, and keep the insect immersed in saline. Trace the nerves from the brain and ganglion, and draw the nervous system.

In order to observe the endocrine glands (corpora allata and corpora cardiaca) in the adult fly, pin an anesthetized or cooled specimen through the thorax and place it into a dissecting dish. Flood with water or saline. Take another pin and fasten the fly, dorsal side up, through the head. Carefully stretch the head forward to separate it from the thorax. Under a dissecting microscope, remove the membrane of the neck. By carefully removing trachea and other tissues, you should observe a round gland located in front of the proventriculus and on top of the digestive tract. This is the corpora allata (c.a.). Just underneath the c.a. is the hypocerebral ganglion-corpora cardiaca complex. Refer to Dethier (1976) for an excellent account of these glands in adult flies.

C.   Observing the Cockroach's Central Nervous System

Follow the same procedure as above, using an adult cockroach of either sex. Identify the components of the central nervous system. (Read ahead in Part E for help in identifying parts.) Trace the nerves from the brain and ganglion, and prepare a drawing of what you observe, referring to Fig. 28-3 as needed. Compare the nervous system of the adult fly and the adult cockroach.

D.   Observing the Central Nervous and Endocrine Systems of the Fly Larva

The larvae of house flies, blow flies, and flesh flies, like other cyclorrhaphous Diptera, have a ring gland (also referred to as Weismann's ring) formed by the fusion of the endocrine system into 1 unit connected to the brain by a pair of nerves (see Fig. 28-4). Presumably this fused gland performs all the functions normally found in the non-fused prothoracic glands, corpora cardiaca, and corpora allata of other insects.

Select a large living fly larva. Place it on an ice cube to slow down its activity. Position the larva dorsal surface up in a dissecting dish. Pin through the tip of its head and the tip of its posterior near the caudal spiracles. Make sure the pins are secure, since the maggot will make every effort to get away.

Using 2 pair of forceps, grasp the maggot at the midline of the dorsum and tear open the cuticle. The internal organs will rush out. Flood the specimen with saline, and continue tearing the cuticle along the midline while moving forward. (Blow fly and flesh fly larvae possess a crop but house fly larvae lack one.) Lying underneath and associated with an inverted V-shaped group of trachea is the ring gland (Fig. 28-4). This structure is located just in front of the brain hemispheres.

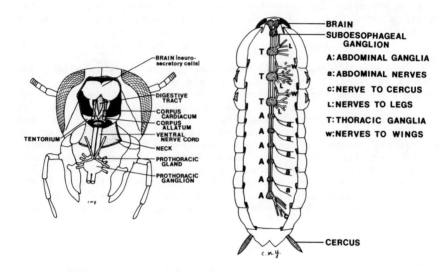

Fig. 28-3. Anatomy of the neuroendocrine system of the American cockroach. A. Posterior view of the head of a nymph showing the principal parts of the neuroendocrine system. B. Dorsal view of an adult showing the paired ganglia of the central nervous system and principal nerves.

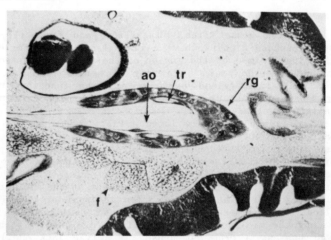

Fig. 28-4. Light micrograph of a house fly larva showing a longitudinal section of the ring gland (rg). Key to abbreviations: ao = aorta; f = fat cells; tr = trachea. With permission from Cantwell, et al. (1976).

Using your forceps under the dissecting microscope, carefully remove the complete ring gland. Place a drop of saline on a clean microscope slide and orient the ring gland in the drop. Place a thin rim of petroleum jelly on the perimeter of a cover slip and carefully place the cover slip over the drop of water containing the gland. Observe the slide under a phase contract microscope. Draw and label the gland.

After observing the ring gland, return to the dissection to observe the central nervous system of the larva (Fig. 28-5). Note the 2 hemispheres of the brain (Fig. 28-5A) and the fused thoracico-abdominal ganglion (Fig. 28-5A, B). Does the exterior of the larval brain show the same specialized regions as the adult brain? If not, explain.

A sagittal section through the central nervous system of a fly larva shows the various fused regions and the orientation of the neuron cell bodies at the periphery. The axons and dendrites form the central region of the ganglion, called the neuropile (see Fig. 28-6). This unique organization of invertebrate ganglia facilitates intracellular staining, mapping, and electrical monitoring of individual neurons.

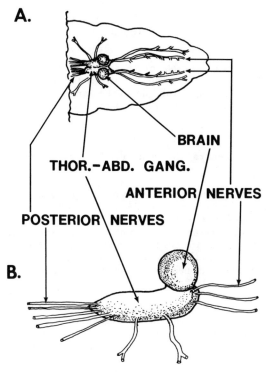

Fig. 28-5. Drawing of the central nervous system of a fly larva. (A) Dorsal view and (B) lateral view showing the brain hemisphere, thoracico-abdominal ganglion and posterior and anterior nerves.

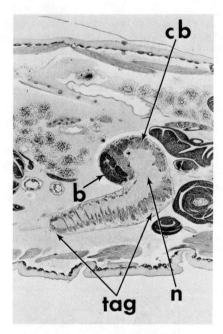

Fig. 28-6. Light micrograph of a sagittal section of a fly larva (see Fig. 28-5B) showing the brain (b) and thoracico-abdominal ganglion (tag), the greyish dendrite and axon-packed area, the neurophile (n) and the outer, darker region containing the cell bodies (cb).

E. Observing the Central Nervous and Endocrine Systems of the Adult Cockroach

     After catching an adult cockroach and anesthetizing it with carbon dioxide, use a razor blade to cut off its head. Save the thorax and abdomen for future dissection by placing the body in a dish or beaker of saline to prevent dessication. In a dissecting dish, place the head in cold tap water with the opening (foramen magnum), where the head joins the neck, oriented upward. Using an insect pin, fasten the head through the mandible area. Carefully remove excess material from the occiput, and pull off the labium (lower lip).
     Look for the white tentorium and the brownish digestive tract (see Fig. 28-3A). Grasp the brown digestive tract with your forceps and move it as though trying to pull it out. (Do not really pull it out.) As it moves, you should note 2 bluish structures. These are the corpora cardiaca.
     Carefully remove any material around these glands. Then insert your forceps around and down into the brain area. Carefully remove this mass of tissue. This should remove the entire corpora cardiaca

and the corpora allata. Draw both pairs of gland, and check your drawing with your instructor.

NOTE: The copora allata are often used in behavior studies involving transplantation; be sure you feel confident about locating them.

After observing the endocrine glands of the adult, return to the dissection. Carefully remove the cuticle from the head, exposing the underlying brain. Locate the protocerebrum, deutocerebrum, tritocerebrum, antennal nerves, optic lobes, and circumesophageal connectives.
Return to the thorax and abdomen of the cockroach you started working with, and make lateral cuts down the sides of the specimen. Remove the dorsal surface, leaving the ventral portion. Pin the specimen into the dissecting dish, flood it with tap water, and carefully remove any tissues or organs until the ventral nerve cord is exposed (see Fig. 28-3B). Notice whether the nerve cord is single or paired. Trace the cercal nerves from the last abdominal ganglion to the cerci.

F.   Observing the Endocrine System of an Immature Cockroach

The prothoracic glands in an insect produce a hormone which initiates growth and development and causes molting. Soon after the insect's final molt to adult, they normally break down. Therefore, in order to observe prothoracic glands we must look in an immature insect.
Anesthetize a live cockroach nymph with carbon dioxide, then inject methylene blue solution into its thorax. Pin the specimen in a dissecting dish, dorsal surface up. Remove the legs. After 2 min., flood the specimen with cold tap water.
In order to locate the prothoracic glands, first look for structures associated with them; these glands are located in the prothorax, are close to the thoracic ganglion, and meet near the ganglion to form an X-shaped structure (Fig. 28-3A). When you have located the prothoracic glands, draw them. Call the instructor for confirmation.

SUMMARY QUESTIONS

1.   Name at least 2 structural differences in the organization of the nervous systems of the cockroach and fly, and relate these to functional differences.

2.   Name at least 2 structural differences in the organization of the nervous systems of the fly larva and adult, and relate these to functional differences.

3.   Why do the corpora cardiaca of freshly dissected insects appear blue?

4. List 3 techniques used by insect neuroanatomists to map the neural circuitry of an insect.

5. List 2 techniques used by insect behaviorists to demonstrate the effect of hormones on the initiation and expression of overt behavior.

6. Prior to demonstrating the role of neural feedback in the cessation of feeding in the blow fly, Phormia regina, Dethier and Gelperin had to locate 2 nerve tracts in the adult fly. What nerve tracts were these? How did Dethier and Gelperin demonstrate the involvement of these tracts in feeding? (Refer to Exercise 14.)

SELECTED REFERENCES

Bullock, T.H. and G.A. Horridge. 1965. Structure and Function in the Nervous Systems of Invertebrates. vols. I, II. W.H. Freeman and Co., San Francisco.

Cantwell, G.E., A.J. Nappi and J.G. Stoffolano, Jr. 1976. Embryonic and postembryonic development of the house fly (Musca domestica L.). U.S.D.A. Tech. Bull. No. 1519. Washington, D.C. 69 pp.

Chapman, R.F. 1969. The Insects: Structure and Function. American Elsevier, New York.

Cornwell, P.B. 1968. The Cockroach. vol. I. Hutchinson, London.

Dethier, V.G. 1976. The Hungry Fly. Harvard Univ. Press, Cambridge, Mass.

Galtsoff, P.S., F.E. Lutz, P.S. Welch and J.G. Needham (eds.). 1959. Culture Methods for Invertebrate Animals. Dover, New York.

Hoyle, G. 1970. Cellular mechanisms underlying behavior - neuro-ethology. Adv. Insect Physiol. 7: 349-444.

Huber, F. 1974. Neural integration (central nervous system). In The Physiology of Insecta. M. Rockstein (ed.), 2nd ed. vol. IV, Academic Press, New York, pp. 3-100.

Kater, S.B. and C. Nicholson (eds.). 1973. Intracellular Staining in Neurobiology. Springer-Verlag, New York.

Lane, N.J. 1974. The organization of insect nervous systems. In Insect Neurobiology. J.E. Treherne (ed.), American Elsevier, New York, pp. 1-71.

Miller, T.A. 1979. Insect Neurophysiological Techniques. Springer-Verlag, New York.

Roeder, K. 1967. Nerve Cells and Insect Behavior. revised ed. Harvard Univ. Press, Cambridge, Mass.

Strausfeld, N.J. and T.A. Miller (eds.). 1979. Neuroanatomical Techniques. Springer-Verlag, New York.

Treherne, J.E. (ed.). 1974. Insect Neurobiology. American Elsevier, New York.

Truman, J.W. and L.M. Riddiford. 1974. Hormonal mechanisms underlying insect behavior. In Advances in Insect Physiology. J.E. Treherne, M.J. Berridge and V.B. Wigglesworth (eds.), Academic Press, New York, pp. 297-352.

29. Grooming in the Fly *Phormia*

William G. Eberhard
Smithsonian Tropical Research Institute
and Universidad de Costa Rica

Janice R. Matthews
University of Georgia

When an entomologist has the opportunity to observe insects alive under a dissecting microscope, he is usually struck by how beautiful and clean they are compared to the specimens he is used to seeing in collections. This is no accident; many insects spend substantial amounts of time cleaning themselves. Such behavior is called grooming. Its selective advantages seem obvious: sense organs are kept in readiness to receive stimuli, and locomotor organs are maintained in condition to move the organism quickly and effectively. The movements that animals use to clean themselves vary between species, and are often quite stereotyped within a species as to pattern and/or sequence. The form and pattern of movements have even been used as taxonomic characters, just like morphological features, to classify some insects.

The objectives of this exercise are to study the cleaning movements of a fly, and to observe changes resulting from different treatments so as to begin to understand the mechanisms controlling this behavior. First you will familiarize yourself with the behavior patterns you will be studying. Then you will observe a series of 3 individuals under 3 different experimental conditions.

METHODS

Subject

Phormia, a common blow fly (Calliphoridae) which breeds in carrion in the wild but is easily reared in the laboratory, has been used successfully in this exercise. Its relatively large size, reliable performance, and availability make it preferred. However, Sarcophaga and Calliphora are also readily available from biological supply companies or field collections; either of these can probably be substituted.

Materials Needed

Three flies in individual vials per student; single edged razor blades or fine scissors (preferred, about 1 pair for every 6 students); stopwatches or wrist watches with second hands; refrigerator or ice baths (preferred); 1 extra vial and lid per student; cornstarch; grease pencils for labeling vial lids; dissecting microscope (optional).

Procedure

A. Manipulating the Flies

NOTE: Attention to the following procedural points is imperative if you are to complete this exercise successfully!

1. There is substantial variation in the behavior of different flies, so you must use each individual for all 3 treatments, and keep the data for that individual together. This requires careful labeling and record keeping.
2. Transfers of non-chilled flies between vials are best done by making use of a fly's strong positive phototaxis. Keeping the bottom end of the vial toward the light, with 1 hand hold the container the fly is in. Open the lid, then place the mouths of the fly-containing vial and the new empty vial together. Turn the vials 180°, so the bottom of the new vial is now toward the light. The fly will usually immediately enter the second vial. Pay close attention to this detail. Otherwise a treasured experimental animal may take off into the wild blue!

3. Flies are disturbed by nearby movements (such as pencils writing notes) and by substrate vibrations. Eliminating these stimuli will make results of different treatments and diffeent individuals more comparable. Rig up some sort of situation where you can write without being observed by the fly. (See Fig. 29-1 for an example.) If you hold your head relatively still, you can move quite close without disturbing the fly. To minimize vibrations, avoid bumping the table, dropping objects on it, etc.

B. Developing a Catalog of Grooming Acts

The first step in any behavioral study is to identify and define the acts of interest, in this case the grooming movements. (See also Exercise 3.)

Obtain 2 vials with flies which have been kept at room temperature for the last few hours. Label the vial lids with a code to identify your individuals, and place the vials behind the viewing screen. Observe your flies for 15 min or so, noting the different acts they use to clean themselves. Attempt to record the duration of different grooming acts in addition to frequency and sequence.

How many categories of grooming can you recognize? The decision of how many discrete states a continuous stream of behavior divides into is, of course, an arbitrary one. In practice, however, it is best to attempt to make a mutually exclusive list of behavioral acts. This means that at any moment the fly has to be doing 1 of them, and

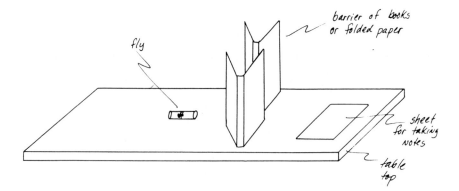

Fig. 29-1. A possible set-up for recording a fly's behavior without disturbing it. Many alternative arrangements will work.

only 1. Invent a code for yourself for each grooming act (such as "1-1" for "leg 1 rubs leg 1", "1-e" for "leg 1 brushes eye", etc.) Do not attempt to distinguish individual movements in an unbroken sequence of a behavior. Flies usually perform a series of acts, rest or move about, then perform another series. Each series should be designated as 1 "bout". You should note the starting and finishing time for each bout.

After you have familiarized yourself with the fly's grooming movements, make a data sheet or checklist to simplify data collection. Practice with it. (You will have to concentrate to keep up with the fly!) When you feel comfortable with your system (hopefully within 15 to 30 min of starting), return your 2 warm flies to the ice bath.

NOTE: Be sure these and all vials stay upright, with their lids out of the ice bath. Water leaking into the vials will ruin your experiment!

C. Normal Post-chilled Fly Grooming

Obtain a warm clean vial and a cold vial containing a chilled fly. Label your vials as they are used. It is imperative that you keep track of individuals and treatments.

Transfer the chilled fly to the warm clean vial. (Place the empty chilling vial back in the ice bath.) Wait 1 min for the fly to warm up, then record its behavior continuously for 10 min, noting each type of movement and the termination time for each bout of behavior. These observations constitute 1 "normal post-chilled" sample.

Return the fly to the chilling vial in the ice bath. Repeat the above procedure with 2 more flies, so you have a total of 3 normal post-chilled samples.

D. Dusty Intact Fly Grooming

While your flies are re-chilling, prepare a warm dusty vial by putting cornstarch into it, closing the lid, and shaking the vial. Remove lid, shake out any excess powder, and recap. Follow the same procedure to re-dust the vial before each use.

For this and the subsequent section, use the flies in the same order as you did for part C. (In this way, you minimize the possible effects of quite different lengths of chilling.)

Remove your first fly and vial from the ice bath. Transfer the fly to the warm dusty vial, and return the empty chilling vial to the ice bath. Wait 1 min for the fly to warm up, then record its behavior for 10 min as before. This is 1 "dusty intact" sample.

Return the fly to the chilling vial. Repeat with the other 2 flies so you have a total of 3 dusty intact samples.

E. Grooming When Dusty Minus Front Tarsi

Remove your first chilled fly from its vial, and place it on its back. (You may wish to put it on the stage of a dissecting microscope, if available.) Immediately -- before the fly warms up! --cut off both front tarsi at the tibia-tarsus junction with a single edged razor blade or fine scissors.

Place the amputee in the warm dusty vial once again. Wait 1 min for it to warm up, and record its behavior for 10 min as before. This is 1 "dusty amputee" sample. Return the fly in its chiling vial to your instructor. Repeat with the other 2 flies, to give a total of 3 dusty amputee samples.

F. Data Analysis

Do the flies groom differently after the experimental treatments? This question might be approached in several different ways. Your data fall into at least 3 categories: (1) data on frequency of the various grooming acts, as defined by your behavioral code; (2) data on duration of individual grooming bouts (which may include more than 1 act); and (3) data on sequence of grooming acts.

Different statistical approaches can be used to measure "differences" in grooming, and the rigor and extent of statistical analysis will depend on your instructor's wishes. Here are some suggestions; see Appendix 1 for methodology.

1. Suppose you make the assumption that all the flies used were randomly selected from the same large population and have similar pre-treatment backgrounds. Then it would be appropriate for each experimental treatment to lump the data for your 3 experimental animals, or indeed for all the flies in the class. You could compute the mean values for each treatment and compare them, perhaps by constructing a graph or histogram. Comparing the mean and median

values would demonstrate whether the sample data are normally distributed. Computing the standard deviation and standard error of the mean would give a measure of the variability in the data (see Exercise 2). A chi square test or Mann-Whitney test could be used to compare samples.

2. It is likely that you have an intuitive feeling that individual fly behavior varied widely. Some animals perform more vigorously, some less, when are exposed to the same environmental situation. Perhaps this variability (the source of which is unknown) is obscuring differences resulting from the experimental treatments -- differences that might show up if you were to trace the data for each fly individually. In fact, there is a straightforward statistical test designed to do just that -- the Wilcoxon Test (Appendix 1.3). The Wilcoxon Test uses information about both the direction of the differences within pairs of samples and the relative magnitudes of the differences. More weight is given to a pair which shows a large difference between the 2 conditions than to a pair which shows a small difference. With this test, you can ask whether differences in frequency or duration of grooming acts or bouts are statistically different between the treatments. You will need to use data from all or part of the class to obtain an appropriate sample size.

3. Data on sequences of grooming acts require another type of analysis. See Suggestions for Further Study, part 1, if you wish to pursue this.

G. Discussion

Your experimental procedure and technique were almost certainly not perfect. Describe the drawbacks and limitations you noticed. Then discuss how they may have affected the confidence you have in the conclusions you reached on the basis of statistical analysis. This last step is important because, paradoxical as it may seem, errors do not always invalidate results. Consider this hypothetical example. If you found in your data that dusty intact flies tended to perform longer bouts of grooming, and you think that you probably erred by underestimating the durations of dusty intact bouts (because, for instance, they occurred so frequently with cornstarch that you couldn't note the time quickly enough), then your error makes you _more_ confident in the trend indicated by your data.

SUGGESTIONS FOR FURTHER STUDY

1. Sequences of Grooming

In observing your flies and/or inspecting your grooming data, you may have noted a strong tendency for certain acts to alternate with others. Grooming seems to divide into discrete bouts that have sudden beginnings and endings, and that seem never to mix front and rear end movements. The order in which specific grooming behaviors occurs in blow flies has been the subject of a study by Dawkins and Dawkins (1976); refer to their paper for an interesting analysis and discussion of this topic and compare their results with yours. Other relevant grooming studies include Zack's (1978a,b) work on praying mantids and Lefevbre's (1981) study of cricket grooming.

2. Response to Different Contaminants

Humans find some kinds of contamination more bothersome than others. Do flies also? Devise an experimental situation in which the flies walk on walls of vials or petri dishes in which aqueous solutions of different substances with different odors have evaporated. Measure responses, and perform appropriate statistical tests upon your data.

3. Response to Partial Dusting

Do flies respond in a directed manner to dust on part of their bodies? One technique might be to cool some flies down, apply cornstarch to selected parts of their bodies with careful strokes of a fine brush, then observe them in clean vials. Again, statistical treatment of results will be necessary.

4. Response to Chilling

Two variables have actually been introduced in this exercise -- the effects of dusting and the effects of chilling. It is possible that the results of chilling have materially affected the responses you observed with dusting. Design and carry out 1 or more experiments to isolate the effects of these variables, and discuss your results. You have already made some observations that could serve to test whether duration of chilling previous to observation period makes a difference. Describe how you could use the data from the entire class to examine this.

5. Comparative Studies of Fly Grooming

The easy availability of various genera of flies, and the comparative stereotypy of their grooming movements, could form the basis for an interesting comparative study. Farish (1972) has done a similar study in the Hymenoptera.

SUMMARY QUESTIONS

1. Did the presence of cornstarch induce any changes in the flies' behavior? What were they? Do the changes seem adaptive for the fly? Explain.

2. Did amputation of the front tarsi change the flies' behavior? What were the changes? Propose at least 5 plausible mechanisms of control of grooming which would account for the observed effects. What experiments might distinguish between these possibilities?

3. Flies clean themselves even when they are in very clean environments. An internal mechanism which spontaneously stimulates them to clean periodically has been proposed. Develop a model of

your own for the control of grooming behavior which incorporates this mechanism and is in agreement with your own results. Describe an experiment (other than the one you performed) which would test whether your model has validity.

SELECTED REFERENCES

Dawkins, R. and M. Dawkins. 1976. Hierarchical organisation and postural facilitation: Rules for grooming in flies. Anim. Behav. 24: 739-755.

Dethier, V.G. 1962. To Know A Fly. Holden-Day, San Francisco.

Farish, D. 1972. The evolutionary implications of qualitative variations in the grooming behavior of the Hymenoptera (Insecta). Anim. Behav. 20: 662-676.

Lefebvre, L. 1981. Grooming in crickets: Timing and hierarchical organization. Anim. Behav. 29: 973-984.

Oldroyd, H. 1964. The Natural History Of Flies. W.W. Norton, New York.

Siegel, S. 1956. Nonparametric Statistics for the Behavioral Sciences. McGraw-Hill, New York.

Zack, S. 1978a. Head grooming behaviour in the praying mantis. Anim. Behav. 26: 1107-1119.

Zack, S. 1978b. The effects of foreleg amputation on head grooming behavior in the praying mantis Sphodromantis lineola. J. Comp. Physiol. 125: 253-258.

30. Laterality in *Drosophila*

Lee Ehrman
S.U.N.Y., Purchase

Ira B. Perelle
Mercy College

The preference for the use of a paw, claw, or hand on one side of the body rather than the other appears in many animal species. Asymmetrical lateral biases appear to exist both as morphological and as behavioral traits. A typical example of the morphological expression of asymmetry is the pair of dimorphic claws of the lobster, Homarus americanus, a cutter and a crusher. Behavioral expression of asymmetry has been observed in chimpanzees, parrots, rats, monkeys, cats, and humans. Human asymmetry is probably a function of morphological asymmetry, the asymmetrical development of the brain.

Several asymmetrical lateralized behaviors can be investigated in Drosophila. Purnell and Thompson (1973) examined genetic components of lateral asymmetry in D. melanogaster. Beginning with a wildtype strain and continuing for 15 generations, 2 lines were selected: left wing folded over right, and right wing folded over left. A small asymmetrical bias was initially achieved, but was lost after the first few generations. Artificial selection to modify bristle and hair number asymmetries has also been attempted in D. melanogaster (Beardmore, 1965; Reeve, 1960; references in Purnell and Thompson, 1973). In addition to wing folding behavior, Drosophila may be scored for directional choices in mazes and for several laterally classifiable behaviors used in courtship.

Drosophila males employ 3 behavioral traits in courtship that can be distinctly classified as right or left behaviors (Fig. 30-1): circling, tapping, and wing extension. This exercise is designed to reveal any lateral preferences in Drosophila during these courtship behaviors. For additional analyses, the flies may be divided into 2 groups: successfully courting males and unsuccessfully courting males, with success defined as copulation within 30 min of the time the male is introduced to a mature female. In this way, perhaps a lateralized component of successful courtship can be identified. You may also examine laterality of courtship behavior in families of Drosophila siblings to determine whether lateral consistency appears within families.

METHODS

Subject

Fruit flies, Drosophila, are the small brownish insects commonly seen around decaying fruit. Their ease of rearing and rapid reproduction rate have made them ideal for genetic studies, and cultures are readily available from a wide variety of sources. For the investigation of laterality in Drosophila, use males and females from any of the following species or strains: D. melanogaster, mutants or wildtype; D. simulans; D. immigrans; D. pseudoobscura; D. persimilis; D. willistoni species group (D. willistoni, D. paulistorum, D. tropicalis, D. equinoxialis, or D. insularis).

Materials Needed

Stock fruit fly culture; anesthetization apparatus; aspirators; transparent observation chamber (Fig. 30-2); binocular dissecting microscope with cool light source, or 4x hand lens.

Procedure

A. Courtship Laterality of Drosophila Males

Your instructor will have previously separated virgin males and females as outlined in the Aids to the Instructor. For observation of courtship behavior and scoring of behavior direction, 1 male and 1

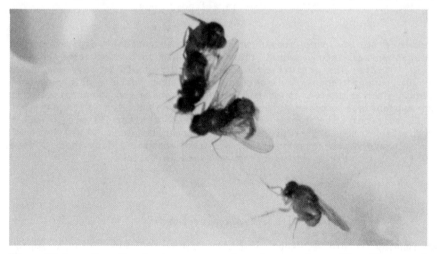

Fig. 30-1. Two D. immigrans couples in copula. The fly on the right is a male, attracted to and orienting toward the couples. Mounts will be maintained for up to an hour; the preceding courtships, however, were but seconds long. (These flies were freshly collected by Prof. Peter A. Parsons in Australia; photographed by Diane A. Granville.)

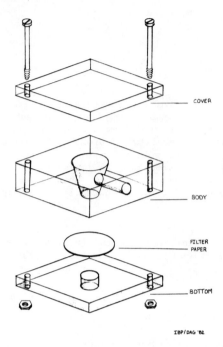

IBP/DAG '82

Fig. 30-2. Observation chamber for <u>Drosophila</u> behavior studies. A small version of the Elens-Wattiau chamber used by Ehrman (1965), it allows observation of the flies in all positions.

female from the same strain should be introduced by aspiration (do not etherize) into a small transparent plastic observation chamber (Fig. 30-2). The activities of each couple should be observed under a low power binocular dissecting microscope, with continuous cool light, over a period of 30 min or until copulation takes place. If an appropriate microscope is unavailable, a hand lens with daylight will suffice. The female should be removed after 30 min or after copulation. Repeat the process with a fresh female.

To score your observations, use a standardized data sheet (Table 30-1). Each male's behavior can be coded as follows:

    0   circling
    +   tapping
    -    wing extension
    __   end of bout with a particular female

A running record is kept down the narrow columns. For example,

    RO
    R-
    R-
    L+

    __

TABLE 30-1. Data sheet for scoring <u>Drosophila</u> laterality.

Scorer: Date:

<u>Drosophila</u> species:

Actions:

DATA SUMMARY

	First Female	Second Female		Total Frequency First Female	Second Female
Time of Introduction:			Circling: RO LO		
Copulation Time:			Tapping: R+ L+		
Total Time:			Wing Ext.: R- L-		

indicates that in this bout the male courted the female by making a right circle around her, then extending his right wing twice (returning the wing to the closed resting position before again extending it), then tapping with his left foreleg.

B. Courtship Laterality Within Families

Male siblings from approximately 12 Drosophila females should be used as subjects. Each family should be raised separately after eclosion; subjects should be separated by sex and aged as before, but siblings should be identified throughout the experiment. Observation and scoring are the same as described for single males, but data should be analyzed for families rather than for population samples.

SUGGESTIONS FOR FURTHER STUDY

As an adjunct to this insect laterality exercise, you may wish to examine human laterality in your own class (or within a larger study population). Begin by having your subjects complete the following questionnaire (Table 30-2). Tally your results, and discuss them with the class. What generalizations can you make? What further information might be helpful?

SUMMARY QUESTIONS

There are clearly 4 primary questions that one can ask about the laterality of a behavioral trait. Apply them to your data, and discuss.

1. Is there any laterality preference or lateral bias evident in your sample population?

2. Do individuals show any significant lateral preferences?

3. Is there consistency in the lateralized use of assorted body parts (for example, wings and legs on the same side)?

4. Are the lateral biases familial?

SELECTED REFERENCES

Beardmore, J.A. 1965. A genetic basis for lateral bias. Symposium on the Mutational Process, Prague. Mutation in Populations: 75-83.
Ehrman, L. 1965. Direct observation of sexual isolation between allopatric and between sympatric strains of the different Drosophila paulistorum races. Evolution 19: 459-464.
Ehrman, L. and P.A. Parsons. 1981. The Genetics of Behavior. 2nd ed. McGraw-Hill, New York.
Ehrman, L., J. Thompson, I. Perelle, and B. Hisey. 1978. Some approaches to the question of Drosophila laterality. Genet. Res. (Cambridge) 32: 231-238.

TABLE 30-2. Laterality, A Survey Questionnaire.

1. With reference to "handedness", how do you perceive yourself? Please mark the appropriate scale position:

strongly right- handed	moderately righthanded	ambidextrous	moderately lefthanded	strongly lefthanded

2. Please indicate your "preferred" hand for the following activities:

	right	left	either	not applicable
Writing...				
Cutting with scissors...........................				
Cutting with a knife............................				
Manipulating a screwdriver......................				
Manipulating a hammer...........................				
Eating with a fork..............................				
Using a racquet (tennis, squash, etc.)................................				
Throwing a ball.................................				
Bowling...				
Using a toothbrush..............................				
Striking a match (match hand)................................				
Manipulating a shovel (top hand)..................................				
Unscrewing a jar lid (lid hand)..................................				

3. Which writing posture most closely resembles your own?

4. What is your age? _____ Sex?_____

5. In which country did you live when you were learning to write?
   ___USA   ____other (please specify)_____

6. What type of elementary school did you attend?
   ____public   ____parochial   ____private   other (specify)_____

TABLE 30-2, cont.

7. What is your race? ____black ____native American ____white
____oriental   other (specify) _____

8. What is your national (ethnic) background? _____

9. Were you part of a multiple birth? ____yes ____no. If so, of
what type? ____twin, fraternal ____twin, identical
other (specify): _____

10. To the best of your recollection, did anyone attempt to "switch"
your primary writing hand when you were learning to write? ____no
____yes   If yes, why? _____

11. Please indicate, to the best of your knowledge, the primary
writing hand of these relatives:

	Right	Left	Unknown
Mother...			
Father...			
Sex: male female			
Sibling # 1...			
Sibling # 2...			
Sibling # 3...			
Sibling # 4...			
Child # 1...			
Child # 2...			
Child # 3...			
Child # 4...			
Spouse..			

Harnard, S., R.W. Doty, L. Goldstein, J. Jaynes, and G. Krauthamer.
1977. _Lateralization in the Nervous System_. Academic Press, New
York.
Koref-Santibanez, S. 1972. Courtship interaction in the semispecies
of _Drosophila paulistorum_. Evolution 26: 108-115.
Perelle, I., S. Saretsky, and L. Ehrman. 1978. Lateral consistency
in _Drosophila_. Anim. Behav. 27: 622-623.
Purnell, D.J. and J.N. Thompson, Jr. 1973. Selection for asymmetrical
bias in a behavioral character of _Drosophila melanogaster_.
Heredity 31: 401-405.
Reeve, E.C.R. 1960. Some genetic tests on asymmetry of sternopleural
chaeta number in _Drosophila_. Genet. Res. (Cambridge) 1:151-172.

31. Candles, Moths, and Ants: The Dielectric Waveguide Theory of Insect Olfaction

Philip S. Callahan
U.S.D.A. Agricultural Research,
Science and Education Administration,
Gainesville

Many insects are attracted to artificial light. That insects are attracted to various chemicals is also well known. Most entomologists believe that these responses are mediated totally by vision in the former case, and by a traditional olfactory system in the latter case. An alternative possibility deserves attention, however. Briefly stated, it is that coded, nonlinear <u>infrared</u> (<u>ir</u>) <u>emissions</u> from free-floating, or thin, layers of organic semiochemicals* mediate both attractions.

By their form, arrangement, and dielectric properties, tiny spines (sensilla) on the antennae appear ideally suited to act as dielectric waveguides or resonators to electromagnetic energy. Although there is no mathematical, electrophysiological, or waveguide evidence that either the human or the insect eye responds to far infrared wavelengths, such evidence for the antennal sensilla of insects is considerable (Callahan, 1965, 1967, 1969, 1975a, 1975b, 1977a, 1977b).

Such electromagnetic energy has been shown to be given off by at least 1 insect pheromone (Callahan, 1977a). It is probably a general characteristic of a variety of insect scents and host plant scents. The difficulty in such study has been the fact that the nonlinear resonant lines from organic semiochemicals are extremely narrow band (less than 0.5 cm-1 bandwidth), and were undetectable until the invention of the Fourier transform spectrophotometer.

In the case of the cabbage looper moth, these narrow band emissions shift to shorter and shorter wavelengths with decreasing concentration and temperature in the water vapor of the night air. A male moth may pick up the long wavelengths close to the female (where the ir concentration is highest and the pheromone on release from the warm-bodied moth warmest), and short wavelengths further away (where the concentration decreases with distance and the pheromone is cooled by the night air).

How does this relate to artificial light such as a candle flame?

---

* Semiochemicals = naturally produced chemicals that carry information between organisms; an "umbrella" term that includes pheromones, allomones, kairomones, etc. (see Nordland et al., 1981).

A light source also emits ir radiation, though this is invisible to human eyes. In the case of a beeswax candle, the lines of nonlinear, resonant ir energy visible on an infrared spectograph are very close to those shown by molecules of the cabbage looper moth's sex phereme as it luminesces under the stimulation of low intensity blue-ultraviolet (uv) night sky.

A light source can also potentially act in another way. It can increase the attractiveness of any free-floating natural scent in the environment, by irradiating it and pumping it to higher energy levels. If the power source of the light is alternating current, it may also modulate the irradiated scents. Thus, a scent under a street lamp, for example, may attract an insect more strongly than the scent in the ordinary night sky.

There are so many frequencies emitted by a candle between 2 μ and 40 μ (Fig. 31-1), and so many insects are attracted to candles, that it is most appealing to postulate that the candle and other flickering hydrocarbon emittors represent powerful coded mimics of sex or host plant scents of innumerable species. This would explain why a male moth can be attracted away from a female to a candle (Fabre, 1949). Not only is the candle, in all likelihood, emitting the same coded lines as the moth sex scent but also any sex scent from the female, released in the room, is being pumped to a much higher energy output in the vicinity of the candle than in the vicinity of the abdominal tip of the emitting female.

This theory also bears, of course, upon traditional ideas of

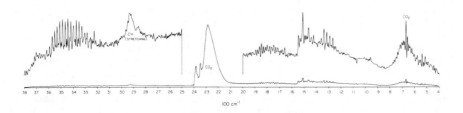

Fig. 31-1. Infrared emission from a 52% beeswax candle, 3800 cm^{-1} (2.63 μm) to 400 cm^{-1} (25 μm). Fourier Transform Spectrophotometer, resolution 2. Emissions scaled in reference to the highest energy peak. The bottom scan shows that in relationship to the total CO_2 energy, (between 2400 cm^{-1} and 2100 cm^{-1}), the hydrocarbon, water, and CH stretching energy (top scan) is very low. The top scan was obtained by rationing out the CO_2 energy centered at 2300 cm^{-1} leaving the rest of the spectrum. At a higher resolution of 0.5 cm^{-1}, many lines would have more than onepeak. The fact that the lines (other than CO_2) are weak has no bearing on insect scent detection since active semiochemical molecules (which the candle emissions mimic) are blowing across the antennae sensilla. What is important is that the narrow band IR emission can couple to the sensilla by resonance.

insect olfaction. For example, Wright (1966) attempted to correlate broad band absorption with olfaction. Although he demonstrated, as would be expected, an excellent correlation, he offered no mechanism for the biological response. Experiments with ant semiochemicals have shown that certain far infrared lines from a candle flame also duplicate emissions from ant semiochemicals in the 20 μ to 60 μ region of the far infrared spectrum. Experiments are underway at present to determine the exact wavelengths and modulation frequencies of such ant semiochemicals.

In this exercise, you will have the opportunity to perform several experiments and to form your own opinions on this controversial theory.

METHODS

Subject

The cabbage looper moth, available commercially, may be used in this exercise at any time of year. However, if seasonality permits, comparing the reactions of various night-flying moths collected at lights will give interesting results. Surgical removal of wings makes the exercise "calmer".

Alternatively or additionally, ants may be used. Conomyrma, Formica, and Solenopsis have been shown to respond to infrared wavelengths. Trying a variety of species is recommended.

Materials Needed

Experimental organisms; roll of heavy duty white wrapping paper (or spray cleaner if desk or table tops are suitable for direct use); matches; grease pencils or felt-tipped markers; meter sticks; aluminum foil. For each student or team: petroleum (dinner table) candle in a holder, and/or 52% beeswax (votive altar) candle; 25 cm tall polyethylene plastic shield; 25 cm tall glass mantle; stopwatch or wristwatch with second hand.

Procedure

This exercise must be conducted in a darkened room. Exact procedures for handling your experimental animal will be discussed by your instructor.

Obtain a clean sheet of white wrapping paper (unless your desk top is very light colored and of a texture suitable for an insect's easy walking), a candle, and a vial containing your experimental animal. Place the candle in the center of the paper. (For safety, protect an area below and around the candle with a square of aluminum foil.) Measure and mark a distance of 50 cm from the candle to serve as your release point.

For each trial, replace the paper with a clean sheet, or clean the desk surface with spray cleaner, in order to avoid possible bias due to contamination by any previously laid chemical such as an alarm or trail substance, etc.

Your candle should be close to "insect level", i.e.,

approximately 2.5 cm tall. If necessary, trim it to this height. Replace the candle as necessary during the exercise. Light the candle. BE CAREFUL WITH THE FLAME AND WITH MOLTEN WAX!

A. Determining the Basic Behavioral Responses

As a "trial run", release an insect at the 50 cm mark. Watch its behavior for 5 min, or until it begins to circle the base of the candle. What actions did you observe? Return the insect to the release point on the paper, and watch once more. Is its behavior different in any observable way? How much so? Does using a fresh ("naive") insect for each trial seem important? Why or why not?
Repeat as necessary until you can develop a data collection sheet that you feel fairly comfortable with. Record both response times and types of behavioral acts. If you are working in pairs or teams, plan on 1 person watching continuously during the trial while another records data; switch roles every few trials.

B. Behavioral Responses to a Bare Candle

Replace the white paper surface, remark the release point on it, and release a new individual. Immediately begin observation and timing. Repeat with new organisms for the number of trials specified by your instructor and/or dictated by availability of experimental animals.

C. Effects of Plastic Shielding

Obtain a 25 cm tall polyethylene plastic shield and place it so it surrounds your lighted candle. This shield passes visible, intermediate, and far infrared wavelengths. However, through its chimney effect it causes hydrocarbon molecules from the candle to flow away high above the experimental animal. Following the same procedure as above, observe and record the behavior of insects released 50 cm from the plastic-shielded candle.

D. Effects of a Glass Mantle

Obtain a 25 cm high glass mantle and set it over your candle in place of the plastic shield. The glass mantle passes visible radiation and near infrared out to 1 um. However, it is completely opaque to intermediate and far infrared radiation. Repeat your experimental procedures again with insects released 50 cm from this glass-shielded candle.

E. Statistical Analysis

The rigor and extent of statistical analysis will depend on your instructor's wishes. If your individual sample sizes are large enough, do your analyses individually; otherwise, class data may be lumped. Minimally, you should compute mean values for each treatment and compare them. Computing standard deviation and standard error of the mean will give a measure of variability in the data. If individual insects were highly variable in their behavior, you may wish to perform additional trials using each individual for all 3 treatments, trace the data for each insect, and perform the Wilcoxon Test (Appendix 1.5).

SUGGESTIONS FOR FURTHER STUDY

1. Behavioral Responses to Different Sources of Light

    In studying insect attraction to artificial light, 1 of the contradictory observations that puzzles proponents of the "visual mediation only" school of thought is that some insects respond in an unexpected manner to different types of light source. One would expect greater attraction to stronger light sources and less to weaker ones. In fact, the variation in response does not quite follow this pattern.

    Devise 1 or more experiments which compare different stationary artificial light sources such as a flashlight, oil lamp, Coleman lantern, mecury vapor street light, ordinary light bulb, etc. Pay attention not only to degree of attraction, but to such behavioral features as focus of the attraction. Use appropriate statistical procedures to help analyze your results.

2. Light and Trail-following of Ants

    The first, and only, reference to ants and candles previous to Callahan's work was in the delightful book, Ants, Bees and Wasps by Sir John Lubbock (1888). Lubbock, of course, was utilizing candles in his work as a source of light for the simple reason that there was no electric light system in existence. He noted that when he moved his candle from 1 side of the ant trail to the other, the ant became confused and reversed direction along the trail. Marak and Wolken (1965) noted the same phenomenon later with fire ants and a 60 watt GE light bulb. In both cases, the behavioral response of the ant species was attributed to the eye alone.

    Devise 1 or more experiments that involve moving an artificial light source in this way. (Might different types of light give different results?) Devise an experiment to test whether sudden changes in natural light give similar results. How might you determine the role of the insect eye in this behavior?

SUMMARY QUESTIONS

1. Most entomologists feel that the attraction of moths to light is totally dependent on the effect of the visible and uv on the moth eye. (It is well documented that the moth eye has peaks of sensitivity in the green and near uv regions.) However, if the attractive force is entirely due to the radiation detectable by the eye, how do you explain your results? Discuss. What experiments might you do next to help resolve this question?

2. The male cabbage looper moth is attracted in a very deliberate fashion to a wax candle; the approach, speed, and deliberate action

of the male moth resemble, in every respect, the approach to a female moth releasing the pheromone scent from the tip of her abdomen. Develop logical arguments for alternative ways in which a candle flame might act as a man-made mimic of an insect's sexual attraction system.

3. How much variability in response did different individuals of a single species display in your experiments? How might you account for this?

SELECTED REFERENCES

Callahan, P.S. 1965. Intermediate and far infrared sensing of nocturnal insects. Part I. Evidences for a far infrared (FIR) electromagnetic theory of communication and sensing in moths and its relationship to the limiting biosphere of the corn earworm, Heliothis zea. Ann. Entomol. Soc. Amer. 58: 727-745.

Callahan, P.S. 1967. Insect molecular bioelectronics: A theoretical and experimental study of insect sensillae as tubular waveguides, with particular emphasis on their dielectric and thermoelectric properties. Misc. Publ. Entomol. Soc. Amer. 5: 313-347.

Callahan, P.S. 1969. The exoskeleton of the corn earworm moth, Heliothis zea, Lepidoptera: Noctuidae, with special reference to the sensilla as polytubular dielectric arrays. Univ. Ga. Agric. Exper. Stn. Res. Bull. 54, Jan. 1967, 105 pp.

Callahan, P.S. 1975a. The insect antenna as a dielectric array for the detection of infrared radiation from molecules. First Internat. Conf. Biomed. Transducers. Paris, France.

Callahan, P.S. 1975b. Insect antennae with special reference to the mechanism of scent detection and the evolution of the sensilla. Internat. J. Morphol. Embryol. 4: 381-430.

Callahan, P.S. 1977a. Moth and candle: The candle flame as a mimic of the coded infrared wavelengths from a moth sex scent (pheromone). Appl. Opt. 16: 3089-3097.

Callahan, P.S. 1977b. Tapping modulation of the far infrared (17 μ region) emission from the cabbage looper moth pheromone (sex scent). Appl. Opt. 16: 3098-3102.

Callahan, P.S., E.F. Taschenberg, and T. Carlysle. 1968. The scape and pedicel dome sensors -- a dielectric aerial waveguide on the antennae of the night-flying moths. Ann. Entomol. Soc. Amer. 61: 934-937.

Fabre, J.H. The great peacock moth. In E.W. Teale (ed.) 1964, The Insect World of J. Henri Fabre. 191 pp.

Lubbock, Sir John 1888. Ants, Bees and Wasps. D. Appleton and Co., New York.

Marak, G.E. and J.J. Wolken. 1965. An action spectrum for the fire ant. Nature 205: 1328-1329.

Nordlund, D.A., R.L. Jones, and W.J. Lewis. (eds.) 1981. Semiochemicals. Their Role in Pest Control. John Wiley and Sons, New York.

Wright, R.H. 1966. Odour and molecular vibration. Nature 209: 571-573.

32. Modeling of Random Search Paths and Foraging Paths

Rudolf Jander
University of Kansas

Local search and foraging paths of animals, such as those studied in Exercises 7 and 12, are among the most random, that is, unpredictable, of all behaviors. This fact exposes the investigatory human mind to an unexpected challenge, because our natural and naive intuition is surprisingly maladapted to the understanding of random processes. For instance, picture in your mind a honey bee foraging fully randomly in a field of uniformly distributed flowers. Obviously and truly, you would expect the bee to have covered a certain distance (D) away from an arbitrarily chosen starting flower (F_o) after visiting a certain number of flowers, say 10 of them. But what is the expected overall linear distance away from the starting flower (F_o) once the bee arrives at the twentieth flower (F_{20})? Is it twice the distance as compared to that covered at the tenth flower? That is, $D_{20} = 2D_{10}$? This naive intuitive expectation is truly false! Therefore, a central purpose of this modeling exercise is to set our naive intuitive reasoning straight, so that it better squares with the realities of random movements in space.

Ultimately, such modeling not only serves our intuition but, more importantly, constitutes the first step toward a more refined mathematical description and analysis of random search paths. This more exacting goal cannot be approached with the following simple exercises, but a few hints may help those who are more ambitious. In the past, all attempts at an adequate mathematical representation of highly random movements of animals proved to be frustratingly difficult. Historically, a breakthrough was achieved early in this century when Albert Einstein clarified many mathematical properties of Brownian motion, which is a fully random movement in space. (Animals, however, rarely move as randomly as Brownian particles do.) Only recently, a mathematical technique developed by Mandelbrot (1977) brought another breakthrough: "fractal analysis", now the proper tool for mathematically describing and analysing random search paths (Jander, 1982). For a more refined application of the simulation model described below, a paper by Pyke (1978) also should be consulted.

The central concept for this exercise is that of the "random walk", which can also be called a "random flight" when specifically applied to a foraging bee. In this context, "random" refers to the choice of turns made between the translatory elements of the walk, the steps, which can be conceived of as straight-line segments.

METHODS

Materials Needed

Minimal requirements: square-ruled graph paper sheets; graph paper with logarithmic scales along both axes; pencils; source of random sequences of 4 numbers (such as tetrahedral dice, pairs of coins, tables of random numbers or (Appendix 2) a calculator programmed to produce random numbers). For the more advanced activities, a programmable calculator or a computer is required. Fig. 32-1 shows a program that can be implemented on the calculators "TI 58" or "TI 59" (Texas Instruments).

Procedure

A. Graphically Describing a Simulated Search Path

In order to simulate the random foraging path of a bee, one has to simplify radically. In this exercise, this simplification concerns both the distances covered between flowers (which we call steps), and the angles (turns) subtended by the approach and departure directions relative to particular flowers.

Our simplified "flower field" is a rectangular grid or lattice with a "flower" at each intersection. Fig. 32-2 illustrates a central portion of such a grid. As further details and definitions are explained, this figure should be consulted continually. The locations of all "flowers" in the grid are defined by rectangular coordinates. The origin of the coordinates, located in the center of the grid, is always the starting point (F_0) for the simulated foraging path. The horizontal axis is the x-axis, with positive values to the right and negative to the left side. The y-axis runs vertically. All straight distances between adjacent "flowers" in the x-direction and in the y-direction are of unit length. That is, the path of the "bee" on this grid mimics a city-block walk, always following 1 or the other axis. "Flowers" are never jumped. Hence, the step length in the simulated random flight is invariably 1.

Due to these constraints, the "bee" has the option at each "flower" to carry out 1 of only 4 turns, which we label by numbers: straight ahead (0), 90° to the left (1), backwards (2), and 90° to the right (3). Note that these discrete angles are labeled in counter-clockwise sequence, as is done in trigonometry, which we have to apply later. It will prove convenient to label the absolute 4 cardinal directions on the grid in a manner similar to the turns, with the positive x-direction defined as "0".

Given the notations as defined, one has a convenient shorthand to describe any path on the grid. All that has to be signified is the sequence of directions taken by sequential steps, because all steps are uniformly of length 1. The direction of the first step has to be defined as an absolute grid direction. All subsequent directions are defined by the turns preceding them. With these stipulations, the path in

000	76	LBL	051	22	INV	101	22	INV
001	11	A	052	77	GE	102	37	P/R
002	36	PGM	053	00	00	103	68	NOP
003	15	15	054	58	58	104	32	X⪦T
004	71	SBR	055	75	-	105	91	R/S
005	88	DMS	056	04	4	106	98	ADV
006	68	NOP	057	95	=	107	25	CLR
007	42	STO	058	42	STO	108	61	GTO
008	04	04	059	13	13	109	11	A
009	43	RCL	060	42	STO	110	76	LBL
010	10	10	061	03	03	111	16	A'
011	32	X⪦T	062	66	PAU	112	42	STO
012	43	RCL	063	02	2	113	10	10
013	04	04	064	00	0	114	68	NOP
014	22	INV	065	44	SUM	115	85	+
015	77	GE	066	13	13	116	91	R/S
016	00	00	067	01	1	117	76	LBL
017	42	42	068	74	SM*	118	17	B'
018	43	RCL	069	13	13	119	68	NOP
019	11	11	070	97	DSZ	120	95	=
020	32	X⪦T	071	00	00	121	42	STO
021	43	RCL	072	00	00	122	11	11
022	04	04	073	00	00	123	85	+
023	22	INV	074	43	RCL	124	91	R/S
024	77	GE	075	01	01	125	76	LBL
025	00	00	076	91	R/S	126	18	C'
026	40	40	077	42	STO	127	68	NOP
027	43	RCL	078	00	00	128	95	=
028	12	12	079	65	x	129	42	STO
029	32	X⪦T	080	02	2	130	12	12
030	43	RCL	081	95	=	131	91	R/S
031	04	04	082	42	STO	132	76	LBL
032	22	INV	083	01	01	133	10	E'
033	77	GE	084	25	CLR	134	42	STO
034	00	00	085	48	EXC	135	09	09
035	38	38	086	22	22	136	68	NOP
036	01	1	087	22	INV	137	91	R/S
037	85	+	088	44	SUM	138	76	LBL
038	01	1	089	20	20	139	15	E
039	85	+	090	25	CLR	140	47	CMS
040	01	1	091	48	EXC	141	02	2
041	85	+	092	23	23	142	42	STO
042	43	RCL	093	22	INV	143	00	00
043	03	03	094	44	SUM	144	42	STO
044	95	=	095	21	21	145	01	01
045	42	STO	096	43	RCL	146	25	CLR
046	02	02	097	20	20	147	58	FIX
047	04	4	098	32	X⪦T	148	02	02
048	32	X⪦T	099	43	RCL	149	91	R/S
049	43	RCL	100	21	21			
050	02	02						

Fig. 32-1. Listing of the calculator program. The notations are those used in programming the TI 58 or TI 59 of Texas Instruments.

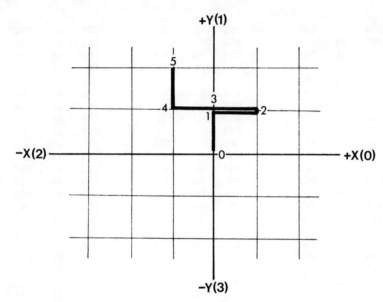

Fig. 32-2. Central portion of the grid used to graph random paths. "Flowers" are the intersections. Numbers inside the grid refer to the consecutively visited "flower" by a sample path marked by a heavy line. The 4 cardinal directions are labeled conventionally and in parenthesis with the number system that is used in all the simulation exercises and in the calculator program.

Fig. 32-2 is unambiguously represented by the sequence of numbers "1, 3, 2, 0, 3."
 As soon as all the notations and definitions are well understood, you are ready for the activities, which will be of 2 types, graphical and numerical ones. We begin with the former.

B. Graphically Constructing An Isotropic Random Path

 For the first simulation you will need a large sheet of square-ruled paper, with the x-axis and the y-axis marked as in Fig. 32-2. In addition, you need a random number source -- a generator producing the 4 numbers 0, 1, 2, and 3 randomly and with equal average frequency (see Appendix 2). For this purpose you could flip 2 coins at a time, thus randomly generating doublets of heads (H) and tails (T). Such doublets can be easily translated into the desired numbers by a rule such as this one:

 HH = 0   HT = 1   TH = 2   TT = 3

This works, but is tedious. Other possibilities include throwing a tetrahedral dice with the 4 faces properly marked, using a table of random numbers, or better still, using an appropriately programmed calculator.

You are now ready to construct your first simulated random foraging path, which is of the <u>isotropic</u> type because turns in all 4 cardinal directions of the circle are equally likely. Draw the first random number; say this happens to be a "3". This means that the first step to "flower" 1 (F_1) is in the negative y-direction. If the next number is a "1", this means a turn 90° to the left, and therefore the next step carries the "bee" into the positive x-direction. Every step chosen is marked by a heavy line.

Construct your own path in this manner until 100-200 steps have been accumulated. As you proceed, consecutively number every 10th "flower" for later use. · The path you have generated and marked by a line gives a good visual first impression of an isotropic random path. It is, in fact, not strikingly different from the utterly erratic path of a Brownian particle.

C. Mathematically Describing An Isotropic Random Flight

Having satisfied our visual curiosity, we are ready for a more rigorous procedure of extracting some principle of order from the disorderly path. Measure the "bee line" distances (D) between the point of origin (F_o) and every 10th "flower" that you marked along your path. (To further clarify the measure D, note that in Fig. 32-2, the values for D are 1,4; 1,0; and 2,2 for the 2nd, 3rd, and 5th flower, respectively.) The outcome of all the measurements are pairs of specified variables with the number of steps (S) selected being the independent variable and the associated distances (D) being the dependent variable. With this data base, one can now ask the crucial question alluded to in the introduction: What is the functional relationship (f) between these two variables? In brief, of what form is D = (f)S?

To find the answer, plot the data points on normal graph paper with the number of steps (S) arranged along the horizontal axis and the distances (D) along the vertical axis. Then, starting at the origin of the coordinates, fit a smooth curve by sight through the scatter of the data points. Typically, this curve turns out to be curvilinear with an upward convexity. This means, in more concrete terms, that the longer the path the more steps (or time) it takes on the average to increase the distance (D) away from the point of origin by a certain amount. Since this thoroughly counter-intuitive outcome is so emminently important, let us study it with more rigor and learn some more details about it.

Our goal now is to put the functional relationship between the distances (D) and the steps (S) into more exact mathematical terms. Fortunately, this can be achieved with adequate approximation by the following simple form of a power function:

$$D = kS^p$$

In this equation, D and S are the already defined variables; k is a multiplicative constant, which we will ignore henceforth, concentrating on the important exponent p, the numerical value of which we plan to extract from the simulation data. To understand how this can be done conveniently, we logarithmize both sides of the above equation and get:

$$\log D = \log k + p \log S$$

This transformed equation describes a straight line, the slope of which is equal to the exponent p.

To apply the above outcome to your data points, replot them on graph paper which has both coordinates logarithmically scaled, thereby repeating the logarithmic transformation applied to the power function. Through this new set of data points, draw the best fitting straight line. The slope p of this line has the same value as the exponent which you want to determine. To get this value, measure the angle between the x-axis and the fitted line. The exponent p is then equal to the <u>tangent</u> (ratio of side opposite to side adjacent) of this angle.*

D. Simulating A Random Flight With Persistence

But are random foraging paths and random search paths really isotropic? Only rarely. A more complex and realistic simulation model would take lack of isotropy into account. The most common and striking deviation from isotropy which one observes in such paths (as in Exercise 12) is <u>persistence</u>, a forward-going tendency. In terms of our simulation model, this means that the 0-turn is more frequently selected than just 25% as in the isotropic example.

In this context, the task is to find out how persistence affects the size of the exponent in the power function formulated above. In other words, how does persistence affect the power law? (One would guess that the exponent will grow, because persistence, when carried to its absolute extreme, will result in a straight-line path, which obviously means an exponent of unity.) To answer such questions, we can simulate a random foraging flight with persistence to determine the exponent p of the power function between the defined variables D and S.

This time, bypass the graphing of the actual path. Use the calculator program (Fig. 32-1) to provide distance values (D) for given numbers of steps (S), both calculated from the simulated random walk. The number of steps for which the program calculates distances should increase in a geometric sequence in order to have an equal spacing of the data points after the logarithmic transformation of the values. The exponent will be determined as has been done in the previous activity.

To produce a noticeable effect, select an example of strong persistence. Let the "bee" go straight on (0) in 90% of all departures from "flowers" ($p_0 = 0.9$), turn left (1) or right (3) in 4% of all turn decisions ($p_1 = p_3 = 0.04$), and double back (2) in the remaining 2% ($p_2 = 0.02$). Enter these 4 frequencies or probabilities

---

*For a more elegant and precise determination of this exponent, use a regression analysis of the logarithmized S values and D values, for which a statistics text or a statistician should be consulted.

(p_0, p_1, p_2, p_3) into the calculator in this exact sequence by means of the respective user-defined keys, A', B', C', and D'.

SUGGESTIONS FOR FURTHER STUDY

The 2 activities described should be considered only as examples of a great variety of possible alternatives. Numerical values for persistence can be changed, and graphical and numerical methods can be combined in various ways. The calculator program could also be used to study the effects of side tendencies.

SUMMARY QUESTIONS

1. Define the following terms: search behavior, random walk, isotropic random walk, random walk with persistence, computer simulation.

2. In what sense does the square-root law formulated for the isotropic random walk run counter to our naive intuitive expectation?

3. Elaborate on some naturally occurring random movements of insects that might be successfully simulated by random walk models.

4. Speculate about the advantage(s) a random foraging path with persistence may have over an otherwise equal foraging path that is isotropic.

SELECTED REFERENCES

Jander, R. 1982. How random is random search? An application of fractal analysis. In preparation.
Mandelbrot, B.B. 1977. Fractals. Form, Chance, and Dimension. W.H. Freeman, San Francisco.
Pyke, G.H. 1978. Are animals efficient harvesters? Anim. Behav. 26: 241-250.

33. Sound Communication in *Aedes* Mosquitoes: A "Dry Lab" Exercise in Data Interpretation

Janice R. Matthews
University of Georgia

In the traditional laboratory exercise in the biological and physical sciences, a great deal of time, effort, and energy goes into the actual experimentation. Although data interpretation is an acquired skill worthy of practice in its own right, it sometimes seems "tacked on" almost as an afterthought. Furthermore, because of the structure of the academic schedule, laboratory exercises also tend to be limited to activities that can be completed within discrete units of time. Thus, interpretation of data tends to be limited to practice in this situation.

The term "dry lab" is often applied derogatorily to the situation in which one skips the process of actually undergoing the experimental procedure, and guesses at the outcome of a laboratory exercise from the reading alone. Used in this way, it shortcuts involvement, interferes with learning, and, if it becomes habitual, defeats the purpose of laboratory instruction.

In another context, however, the "dry lab" approach is used regularly by nearly every working scientist -- each time one critically reads the work of another scientist published in a journal, or listens to a report of it at a seminar or scientific meeting. The very structure of scientific reports invites this approach -- building up background information, comparing alternative hypotheses, presenting data, and finally arriving at specific conclusions. The scientific community may or may not agree with the reporting scientist's conclusions, and lively controversies sometimes ensue.

In this exercise, a sample of this approach -- practice in data interpretation based on published experimental results -- is given. Hopefully you will carry this approach on as you read other studies, thus continuing to improve your skills in data interpretation.

METHODS

Subject

It is well known that mosquitoes produce an audible hum during flight, which may be so distinctive that some researchers claim they are able to identify the species by this sound alone. Through an ingenious set of careful experiments over a period of years, Roth (1948) established that male mosquitoes are drawn to a female mosquito by the buzzing sounds of her wings as she flies about. However, at the beginning of his study Roth had to contend with a number of other hypotheses which researchers had proposed. For example, one claimed that males recognized females by odor. A second postulated that the female's motion alone was sufficient for male recognition and attraction. A third had suggested that knoblike projections (halteres) behind the female's wings produced attractive "bird-like" sounds both at rest and in flight.

Procedure

A. Roth's Experimental Results

Below are listed some of Roth's observations on the yellow fever mosquito, <u>Aedes</u> <u>aegypti</u>. Study them from the viewpoint of a research scientist familiar with the various hypotheses mentioned above.

1. Once male <u>Aedes</u> <u>aegypti</u> began to copulate, 15-24 hrs after emergence, they remained in a constant mating state throughout the rest of their lives, and would mate repeatedly with the same or different females. However, the same males which readily mated with flying females were indifferent to the freshly killed bodies of these females.

2. Although a resting female might be surrounded by males, with some so near that they touched her, never were any resting females observed to induce a male to mate.

3. No evidence was found that female <u>A</u>. <u>aegypti</u> were attracted to sounds, though they sometimes gave shock reactions to certain frequencies. Males, however, showed their characteristic attraction and mating response to any sound in the range between 300 and 800 vibrations per second, regardless of whether produced by a recording of another mosquito, by an audio oscillator, or by a tuning fork.

4. By means of a fine wire looped about their necks, females were suspended in a cage full of males. As long as a female hung with motionless wings, the males remained indifferent, even though Roth might swing her to and fro. However, as soon as she began to vibrate her wings, the males flying nearby immediately seized her and began to mate.

5. When female mosquitoes with halteres removed were suspended in the cage, males were attracted when they vibrated both wings. However, males ignored wingless females, even those which still had their halteres.

6. Cutting off increasingly greater portions of the females' wings caused the sound to decrease in volume and the note to rise progressively. The sounds produced by the cut wings were different

in pitch from those produced by normal females during flight. Yet males were attracted to, and mated with, females which had only parts of their wings, even stumps, vibrating.

7. After caged female mosquitoes were forced to fly continuously so that males mated with them repeatedly, they tended to resist further mating attempts by these males. Yet the sounds produced by the resisting females during flight continued to attract males.

B. Data Interpretation

After studying the results above, write out your own responses to each of the following questions. Hand these in and/or meet for discussion at the time designated in class.

1. What evidence bears most directly on the hypothesis that female motion alone is sufficient in mosquitoes for male recognition and attraction?

2. What does the evidence indicate about the importance of the "bird-like" sounds that one researcher noted?

3. A structure at the base of the female mosquito's wings was suggested to be the principal sound-producing organ. What evidence best supports this?

4. Some researchers have called the sound produced by the female a "mating call." Does the evidence appear to support such an interpretation? Why or why not?

5. It appears that the male mosquito does not respond to a single fixed pitch, but to frequencies which vary over some range. What is 1 advantage such behavior might have? What is 1 disadvantage?

6. Mature female A. aegypti and freshly emerged males beat their wings at about the same rate, and are pursued by older males. Young, freshly emerged females have a different wing beat rate and are not pursued. What advantage might the mature male's behavior have? By what behavior(s) might young male mosquitoes avoid such pursuit?

7. Does odor appear to play a role in the sexual communication of A. aegypti? Why or why not?

8. From what you have learned here, discuss the probable usefulness of biological control of mosquitoes based upon attraction by sound. Be specific in citing support for your stand. What further information might you wish to have? How might you design 1 or more experiments to provide it?

SUGGESTIONS FOR FURTHER STUDY

1. The "bee language controversy" has become a nearly classic
example of differences in data interpretation arising within the
scientific community. As individuals or teams, study the references
on this subject at the end of the exercise. Outline the experimental
results they present, and the questions they attempt to address.
Attempt to formulate each author's working hypotheses. Write a 1
paragraph abstract of the controversy, in a manner similar to that
presented in the Subject section above.

2. Select a recent research paper in some area of insect behavior,
and develop a "dry lab" exercise of your own. Rather comprehensive
studies, such as might be undertaken as a doctoral dissertation, are
generally easier to use than shorter, more limited research reports.
Good studies for this purpose may be found in journals such as Animal
Behaviour, Behaviour, and Zeitschrift fur Tierpsychologie.

SELECTED REFERENCES

Frisch, K. von. 1971. Bees, Their Vision, Chemical Senses and
    and Language. rev. ed., Cornell Univ. Press, Ithaca, N.Y.
Gould, J. L. 1975. Honey bee recruitment: The dance language
    controversy. Science 189: 685-693.
Gould, J. L. 1976. The dance language controversy. Quart. Rev.
    Biol. 51: 211-244.
Jones, J. C. 1968. The sexual life of a mosquito. Sci. Amer.
    218(4): 108-116. (offprint no. 1106).
Roth, L. M. 1948. A study of mosquito behavior. An experimental
    laboratory study of the sexual behavior of Aedes aegypti.
    Amer. Midl. Natur. 40: 265-352.
Roth, M., L. M. Roth, and T. E. Eisner. 1966. The allure of the
    female mosquito. Nat. Hist. 75: 27-31.
Wenner, A. M. 1971. The Bee Language Controversy: An Experiment
    in Science. Educational Programs Improvement Corp., Boulder,
    Colo.

34. Developing Observational Abilities and Generating Hypotheses: Non-Traditional Training Exercises for Behavioral Scientists

R. Stimson Wilcox
S.U.N.Y., Birmingham

It is unlikely that anyone really understands what animal behavior research is about without actually doing it. Probably the best way to gain self-confidence as a researcher is to venture alone into what for you is the unknown, with a good theoretical and methodological base and the receptivity to learn rapidly from mistakes, accidents, and serendipity. In behavioral research, there is an obvious premium on the ability to observe accurately, with a mind open to alternate possibilities of interpretation and unusual occurrences. While there doesn't seem to be any real substitute for a one-on-one relationship between you and the animal there in front of you, it is also helpful to try other ways of training yourself for behavioral research.

The activities in this exercise are designed (1) to capitalize on your natural pattern-resolving abilities, (2) to increase your awareness of the need to allow yourself time to resolve patterns well, (3) to encourage fresh viewpoints in generating hypotheses and interpreting results, and (4) to help you realize that a wide variety of life experiences can aid you in developing the ability to do specialized research effectively.

METHODS

Subject

Students in the class are the subject of these non-traditional training exercises, particularly for parts B-D. In part A, the subjects are 2-dimensional creatures, thought to be invertebrate in nature and classified as Blips, Shibbles, Sharps, Killiwonks, Snappyloos, and Nippybugs. Their behavioral diversity is assumed to mirror their morphological diversity.

Materials Needed

Textured patches approximately 20 cm diameter (such as non-uniform ground, birdskins, scanning EM photographs, etc.); stereograms featured in <u>American Scientist</u> 62 (1974); stereooptic glasses (see Aids to Instructor); 5 dice.

Procedure

The activities which follow focus upon different aspects of visual pattern resolution. After you have experienced them, the hope is that you will transfer aspects of your experiences, in your own way, into a more effective use of your abilities in actual behavioral research and observation.

A.   The Identification of Defining Attributes

Refer to Fig. 34-1. Each of these sets of Alien Animals is a miniature observation/research problem to solve, wherein you first observe a situation, then generate hypotheses, test them out, and interpret the results, returning to further observations if necessary.

Tackle each group of Alien Animals in sequence. Let your mind relax, and try to pick out the relevant cues. Let your visual pattern-resolving ability come to the fore without forcing it, in addition to generating and testing ideas about various kinds of patterns. Take your time, and afterwards think over the types of patterns you had to perceive in order to solve the various sets.

B.   Visual Pattern Analysis: "Just Looking"

This activity is an intensified experience in visual pattern analysis. Decide upon a small textured area (about 20 cm in diameter) to examine. Probably the best texture to look at is a patch of non-uniform ground; however, almost anything will do. For example, when snow and ice have lessened enthusiasm to go outside for this activity, looking at feather patterns on birdskins has been effective.

Without thinking of anything in particular, look at the textured area for 10 min or more. What different patterns do you see during this time period?

Most people experience 1 or 2 "jumps" in awareness of new types of patterns during the 10 min of "just looking." When we observe animals behaving, "just looking" will similarly result in our seeing new patterns. This is a natural ability we all possess. Insightful questions, experimental designs, and interpretations then follow naturally from such observational absorption.

C.   Observation And The Observer: Stereoscopic Fusion

Obtain the stereosoptic glasses and the stereograms from the instructor. Hold the cardboard glasses so that the green is in front of your left eye. Look at the random-dot pattern in the first stereogram from directly in front of it and from about 45 cm away. Without trying to force it, simply wait for a 3-dimensional figure to emerge. After you have experienced the first figure, try the other stereograms the same way. You may have to be more patient with some of the other stereograms.

236

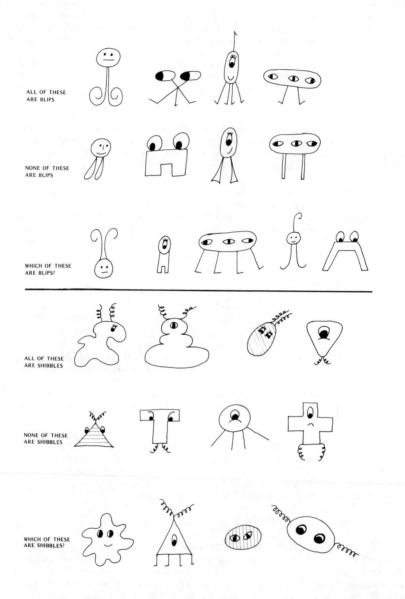

ALL OF THESE
ARE BLIPS

NONE OF THESE
ARE BLIPS

WHICH OF THESE
ARE BLIPS?

ALL OF THESE
ARE SHIBBLES

NONE OF THESE
ARE SHIBBLES

WHICH OF THESE
ARE SHIBBLES?

Fig. 34-1. Alien Animals: Systematic samples of 6 life forms encountered during 21st century intergalactic space travel and brought back to earth for taxonomic study. Can you determine the

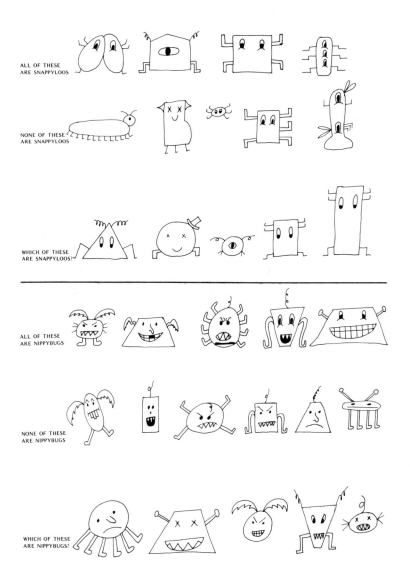

ALL OF THESE ARE SNAPPYLOOS

NONE OF THESE ARE SNAPPYLOOS

WHICH OF THESE ARE SNAPPYLOOS?

ALL OF THESE ARE NIPPYBUGS

NONE OF THESE ARE NIPPYBUGS

WHICH OF THESE ARE NIPPYBUGS?

basis for the scientists' classification? The 6 groups are Blips, Shibbles, Sharps, Killiwonks, Snappyloos, and Nippybugs.

238

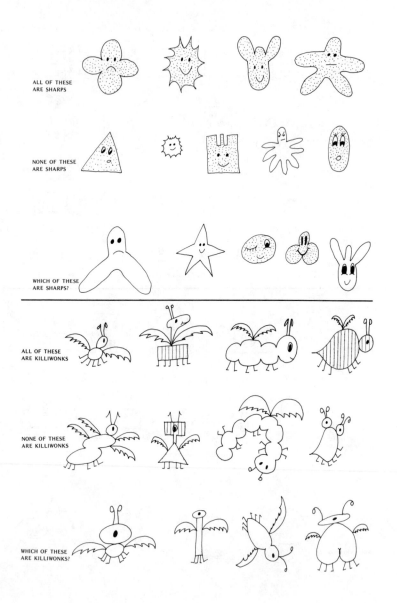

ALL OF THESE
ARE SHARPS

NONE OF THESE
ARE SHARPS

WHICH OF THESE
ARE SHARPS?

ALL OF THESE
ARE KILLIWONKS

NONE OF THESE
ARE KILLIWONKS

WHICH OF THESE
ARE KILLIWONKS?

Fig. 34-1. Continued.

After you have seen the 3-dimensional figures, wait awhile, then try them again. Are you able to see them more quickly the second time?

Obviously, another form of pattern analysis is occurring during this activity. Besides its rather bizarre effect, this activity is meant to emphasize the fact that pattern resolution actually changes us. Most people see the 3-dimensional figures more quickly after the first time, and this tendency holds for years. Simple observation, then, can change us in apparently permanent ways.

Specific interpretations of what changes actually occur during stereoscopic fusion are discussed in Julesz (1974).

D. Formulating And Testing Hypotheses: Petals Around A Rose

This activity is specifically designed to challenge you and hopefully to frustrate you for a while, so that you have to "dream up" a variety of possible hypotheses, test them out in rapid order, then reject unsuccessful hypotheses in favor of new ones. The process is similar to your experiences in the Alien Animals activity. Both emphasize that you should be prepared to let hypotheses fall from your attention when they do not prove fruitful on present evidence. Too many scientists doggedly cling to cherished ideas and hypotheses, thus limiting themselves. We are reminded of the comment that "hope clouds observation."

This game-like activity is to be led by someone who knows the rule or rules. The idea is to discover for yourself what rule(s) determine(s) the number of "petals around a rose" in each throw of the dice.

In groups of between 5 and 10 students, observe the game. When you have figured out the answer, keep quiet so the others will have the opportunity to keep working on the problem. If it takes you longer than others to solve this, be reassured that it simply means your search imagery was inappropriate, not that you are intellectually slow!

The "petals around a rose" game is commonly solved more quickly by people with backgrounds in the humanities (particularly in those involving visual patterns, such as painting, architecture, etc.) than by scientists. Discuss this phenomenon and implications it has regarding the types of mental patterns we are accustomed to thinking in and with which we view the world.

As a group, discuss the variety and progression of ideas you thought up and tried out.

SUGGESTIONS FOR FURTHER STUDY

The Alien Animals pictured in this exercise owe an intellectual debt to activities originally developed for 2 quite different audiences, animal taxonomists and elementary school children. See the Scientific American article on numerical taxonomy (Sokal, 1966) for an engrossing discussion of biological classification which includes a set of imaginary animals, called "caminalcules" after their creator, Joseph H. Camin of the University of Kansas. You may also wish to obtain the Creature Cards which are part of the

Elementary Science Study module, Attribute Games and Problems. These include 9 sets of creatures which vary in identifying attributes and difficulty level. Either caminalcules or Creature Cards could suggest ideas for follow-up activities. For example, try designing your own modifications (such as your own sets of hypothetical animals) or devising completely new types of exercises which emphasize behavioral study principles in a non-traditional way.

SUMMARY QUESTIONS

1. When "just looking," what patterns did you eventually see that were not visible to you at first?

2. Is there an element in common here among all the activities of this exercise? What? In what ways do you think active thought patterns interact with essentially non-thinking pattern resolution abilities? Does "thinking about" patterns influence the patterns you actually see?

3. Ponder each activity in this exercise in turn. How do you think they relate to learning how to do actual behavioral research? When you next watch animal behavior, think about the meaning of these activities at the same time.

SELECTED REFERENCES

Adams, J. L. 1979. Conceptual Blockbusting. 2nd Ed. W. W. Norton & Co., New York.

Elementary Science Study. 1974. Attribute Games and Problems. Webster Division, McGraw-Hill Book Co., New York.

Julesz, G. 1974. Cooperative phenomena in binocular depth perception. Amer. Sci. 62: 32-43.

Sokal, R.R. 1966. Numerical taxonomy. Sci. Amer. 215(6): 106-116. (Dec.)

AIDS TO THE INSTRUCTOR

---

1. CONDUCTING A PERSONAL PROJECT ON INSECT BEHAVIOR. J. ALCOCK

Students often benefit from a chance to put into practice some of the techniques and skills they have acquired in their programmed laboratory exercises. A block of about 4 weeks can be set aside for completion of a project the student had chosen and designed; each student should, however, choose a subject and a plan of action after the first third of the class. This permits an instructor to review each project and to make suggestions, as well as encouraging the student to begin work promptly.

Class members may be asked to give a progress report midway through the block of time set aside for independent projects, with particular emphasis on the difficulties encountered in their work. (The need to make a progress report stimulates progress.) Students can be invited to make suggestions to solve the research problems of their classmates. In the final week of the course, each participant may give a short oral report to the entire class on his or her work before handing in a final written report. A schedule of this sort motivates students to complete their project, and gives them a chance to receive guidance throughout their work. The oral report may be particularly valuable. First, few students have had much experience in public speaking. Second, it promotes quality research, for few people wish to demonstrate to their peers that they have done an inferior piece of work.

2. OBSERVATION, DESCRIPTION, AND QUANTIFICATION OF BEHAVIOR: A STUDY OF PRAYING MANTIDS. E. M. BARROWS

1. Background

Humans probably have always found mantids interesting. Due to their curious praying attitudes, they have been associated with numerous superstitions and religious beliefs; ancient Greeks, for example, called them Mantes, which means prophet. The seemingly tireless insect observer Jean Henri Fabre (1897) and others wrote about the voracious appetites of female mantids and their cannibalization of their mates; Phil and Nellie Rau (1913) described the life history of S. carolina; Przibram (1907), M. religiosa. Didlake (1926) gave methods for rearing T. a. sinensis and S. carolina in captivity. Sexual behaviors of M. religiosa and African mantids were studied by Roeder (1937) and Edmunds (1975), respectively. Other aspects of mantid biology of interest include: locomotion (Roeder, 1937); defensive behavior (Crane, 1952; Maldonado, 1970; Edmunds, 1972, 1976); prey capture (Mittelstaedt, 1957; Copeland and Carlson, 1977); behavioral development (ontogeny) (Balderrama and Maldonado, 1973); egg hatching, grooming (Zack, 1978a,b); color change (James, 1947); ecology (Hurd et al., 1978); foraging (Charnov, 1976); learning (Gelperin, 1968; Maldonado, 1972). Zack (1978b) has published a behavioral catalog (ethogram) and a time budget for the African mantid, Sphodromantis lineola.

2. Obtaining Mantids for Classroom Use

Biocontrol, biological supply, and seed companies sell mantid oothecae. Oothecae also can be easily collected in mantid habitats in autumn after leaf fall. The oothecae can be refrigerated for several months to retard hatching. In the Washington, D.C., area, mantids hatch prematurely during unseasonably warm autumns. If oothecae are brought into a 26°C room in the fall, they hatch at unpredictable times from fall until spring.

Young mantids survive well on a diet of living Drosophila adults; larger mantids feed on crickets of appropriate sizes. In rearing mantids to adulthood in the laboratory, Didlake (1926) maintained individual mantids in clean, closed jars that each contained a strip of filter paper on which mantids clung. Mantids were transferred to larger jars as they grew. In the laboratory, males of S. carolina took an average of 77.5 days to develop from hatching to adulthood; females, 79 days; males of T. a. sinensis, 77 days; females, 78 days. Well-fed, field-collected female mantids will oviposit under laboratory conditions.

In temperate regions, adult mantids can be collected in meadows and along hedgerows in late summer and early autumn. I have collected about 10 males and 10 females of T. a. sinensis in several hours in the same meadow in Reston, VA, in each of the last 5 autumns.

Mantids are often attracted to lights at night, and may stay in illuminated areas to feed on other insects that the lights attract. A 15-minute search around an illuminated auto showroom in Arlington, VA, yielded 10 T. a. sinensis and 4 M. religiosa in September. Students have brought mantids to me after I advertised for them with signs put up around campus. Also, an "action alert" notice in a local newspaper brought in several dozen mantids.

3. Age and Stage of Mantids

Adult mantids are preferable. However, if adults are unavailable, young mantids may be used to observe feeding, grooming, optomotor behavior, defensive behavior, and emergence from oothecae. Magnifying glasses and/or dissecting microscopes may be needed to see early instar mantids adequately.

4. Prelaboratory Preparations

Obtain the materials. Do not feed the mantids for 24 hr before the laboratory exercise is scheduled to begin. Place mantids in the terraria surrounded by arenas in experimental locations at least 1 hr before class.

5. Additional Information on Specific Parts of Exercise

A. Defense

You may wish to amplify discussion of primary defensive adaptations, especially camouflage. Some African mantids can change color on a few days without molting (Rohrmann, pers. comm.). Different species resemble sticks, bark, leaves, and flowers (Edmunds, 1974).

B. Grooming

In female S. lineola, water alone has a probability of 0.42 (N = 200 trials) and 2% acetic acid solution has a probability of 0.80 (N = 50) of releasing head grooming (Zack, 1978a). About 3 ml of liquid were applied to the right or left anterior surface of mantid heads with either Pasteur or precision calibrated automatic pipettes. Mantids were mounted on their backs on inclined plastic blocks and given counterbalanced platforms to grasp with their middle and hind legs during Zack's experimentation.

C. Feeding

Mantids are predators that usually hang from plants and ambush prey. Most species assume a "praying" attitude and remain motionless for hours at a time except for vibrating their antennae (Rilling et al., 1959). Some colorful species attract prey by mimicking flowers (Wickler, 1968). A mantid may stay on the same perch for several days (Krombein, 1963).

As they increase in size through maturation, mantids eat larger arthropods. For example, T. a. sinensis eat aphids and tiny flies as first instars, but as adults eat large insects including adult grasshoppers, crickets, and large butterflies. Some tropical mantids even catch and eat frogs, lizards, and small birds (Ridpath, 1977).

Feeding behavior in mantids consists of assuming an ambush position, visually searching and following prey, sometimes stalking (by very hungry mantids), capturing by an extremely fast grasping movement of the raptorial forelegs, bringing prey to mouthparts, and ingesting (Mittelstaedt, 1957; Rilling et al., 1959; Roeder, 1959). Laboratory work shows that mantids can learn to avoid distasteful, possibly poisonous prey (Gelperin, 1968).

When an object moves into a mantid's visual field, there may be no external reaction, saccadic head movements that follow the prey, or saccadic head movements plus an approach towards prey (Rilling et al., 1959). When the object comes within twice the reach of a mantid's stroke, the saccadic head movements become replaced by smoother head movements (Mittelstaedt, 1957).

D. Ethogram Construction

As a follow-up activity to this exercise, you may wish to have students observe mantid behavior in some detail over a period of time and construct ethograms.

In a total of 75 mantid-hr of observing S. lineola, Zack (1978b) found that mantids were active 31% of the time. Sixty percent of the active time was spent in feeding activities, and grooming comprised 17% of this time. Foreleg grooming represented 82%, and head grooming represented 14% of all grooming.

E. Statistical Analysis

In quantitative analyses of behavior, there has been rapid growth in the application of advanced techniques. Useful books on these quantitative methods include Colgan (1978), Siegel (1956), Sokal and Rohlf (1969), and Zar (1974). The present exercise covers only some elementary methods; more sophisticated techniques, including those which will be developed in the future, undoubtedly will become essential tools for practicing behaviorists.

6. Additional References

In addition to those listed in the student exercise, the following references provide helpful background information:

Balderrama, N. and H. Maldonado. 1973. Ontogeny of the behaviour in the praying mantis. J. Insect Physiol. 19: 319-336.
Charnov, E.L. 1976. Optimal foraging: Attack strategy of a mantid. Amer. Nat. 110: 141-151.

Colgan, P.W. 1978. Quantitative Ethology. John Wiley and Sons. New York.

Copeland, J. and A.D. Carlson. 1977. Prey capture in mantids: Prothoracic tibial flexion reflex. J. Insect Physiol. 23: 1151-1156.

Crane, J. 1952. A comparative study of innate defensive behaviour in Trinidad mantids (Orthoptera, Mantoidea). Zoologica (N.Y.) 37: 259-294.

Didlake, M. 1926. Observations of the life-histories of two species of praying mantis (Orthoptera: Mantidae). Entomol. News 37: 169-174.

Edmunds, M. 1972. Defensive behaviour in Ghanaian praying mantids. Zool. J. Linn. Soc. 51: 1-32.

Edmunds, M. 1975. Courtship, mating, and possible sex pheromones in three species of praying mantis. Entomol. Mon. Mag. 111: 52-57.

Edmunds, M. 1976. The defensive behaviour of Ghanaian praying mantids with a discussion of territoriality. Zool. J. Linn. Soc. 58: 1-37.

Fabre, J.H. 1897. Souvenirs Entomologiques, vol. 5.

Gelperin, A. 1968. Feeding behavior of the praying mantis: A learned modification. Nature 219: 399-400.

Howard, L.O. 1886. The excessive voracity of the female mantis. Science 8: 326.

Hurd, L.E., R.M. Eisenberg and J.O. Washburn. 1978. Effects of experimentally manipulated density on field populations of the Chinese mantis (Tenodera ardifolia sinensis Saussure). Amer. Midl. Nat. 99: 58-64.

James, H.G. 1947. Colour changes in Mantis religiosa L. Can. Entomol. 76: 113-116.

Krombein, K.V. 1963. Behavioral notes on a Floridian mantid, Gonitista grisea (F.) (Orthoptera, Mantidae). Entomol. News 74: 1-2.

Levereault, P. 1938. The morphology of the Carolina mantis. II. The musculature. Univ. Kansas Sc. Bull. 25: 577-633.

Maldonado, H. 1970. The deimatic reaction in the praying mantis Stadmatoptera biocellata Z. vgl. Physiol. 68: 60-71.

Maldonado, H. 1972. A learning process in the praying mantis. Physiol. Behav. 9: 435-445.

Maldonado, H. and J.C. Barros-Pita. 1970. A fovea in the praying mantis eye. I. Estimation of the catching distance. Z. vgl. Physiol. 67: 58-78.

Maldonado, H. and L. Levin. 1967. Distance estimation and the monocular cleaning reflex in the praying mantis. Z. vgl. Physiol. 56: 258-267.

Maldonado, H., L. Levin, and J.C. Barros-Pita. 1967. Hit distance and the predatory strike of the praying mantis. Z. vgl. Physiol. 56: 237-257.

Mittelstaedt, H. 1957. Prey capture in mantids. pp. 57-71 In: Recent Advances in Invertebrate Physiology, B.T. Scheer, Ed., Univ. Oregon, Eugene.

Przibram, H. 1907. Die Lebensgeschichte der Gottesanbeterinnen (Fangheuschrecken). Z. wiss. Insektenbiol. 3: 117-122, 147-153.

246

Ridpath, M.G. 1977. Predation on frogs and small birds by <u>Hierodula werneri</u> (Giglio-Tos) (Mantidae) in tropical Australia. J. Aust. Entomol. Soc. 16: 153-154.
Rilling, S., H. Mittelstaedt and K.D. Roeder. 1959. Prey recognition in the praying mantis. Behaviour 14: 164-184.
Roeder, K.D. 1937. The control of tonus and locomotor activity in the praying mantis (<u>Mantis religiosa</u> L.) J. Exp. Zool. 76: 353-374.
Roeder, K.D. 1959. A physiological approach to the relation between prey and predator. Smithsonian Misc. Coll. 137: 286-306.
Roeder, K.D. 1960. The predatory and display strikes of the praying mantis. Med. Biol. Illustr. 10: 172-178.
Siegel, S. 1956. <u>Nonparametric Statistics for the Behavioral Sciences</u>. McGraw-Hill, New York.
Varley, G.C. 1939. Frightening attitudes and floral simulation in praying mantids. Proc. R. Entomol. Soc. London. Ser. A. 14: 91-96.
Wickler, W. 1968. <u>Mimicry in Plants and Animals</u>. McGraw-Hill, New York.

3. SAMPLING METHODS AND INTEROBSERVER RELIABILITY: COCKROACH GROOMING. R.W. MATTHEWS

The instructor will probably wish to assign initial sampling methods to each team, and rotate teams in the use of the event recorder to expedite wise use of laboratory time. Using a fresh cockroach for each part of the exercise is desirable, but if numbers are limited, roaches may be "recycled". Note that the frequency of grooming diminishes through time after dusting. (This fact could be used as a basis for further study, however.) The laboratory exercise could be used with other organisms and for sampling other behaviors.

If an event recorder is not available, use a tape recorder and stopwatch for Part C. For the data analysis in Part D, convert the running commentary from the taped record onto a continuous graph paper sheet.

For confining roaches to the observation area, Fluon is highly effective, but vaseline is more readily available in reasonable quantities. Fluon may be obtained in 1 gallon lots from Norteast Chemical Company, Woonsocket, R.I. 02895. Specify Fluon (ICI), Grade G.P.-1.

While the data collection is very straightforward, the analysis is time-consuming. Therefore, a division of labor is highly recommended. For additional help in calculations, Lehner's (1978) <u>Handbook of Ethological Methods</u> is suggested. To give you an idea of the format and type of values to be expected, the following hypothetical examples are given. For best results with this exercise, a correlative lecture on sampling methods is suggested.

A. An example of an agreement matrix for some hypothetical data illustrating the calculation of Kappa (from Sackett, 1978).

	BEHAVIOR CODES	Observer 1						PROPORTION OF TOTAL FOR OBSERVER 1 (P_1)
		A	B	C	D	E	F	
	A	26	0	0	0	0	0	26/120 = 0.217
	B	0	30	0	0	0	0	30/120 = 0.25
Observer	C	0	0	18	9	0	0	27/120 = 0.225
2	D	0	0	0	11	0	0	11/120 = 0.092
	E	0	0	0	0	0	13	13/120 = 0.108
	F	0	0	0	0	0	13	13/120 = 0.108

PROPORTION OF TOTAL FOR OBSERVER 2 (P_2)	A	B	C	D	E	F
	0.217	0.25	0.15	0.15	0	0.217

CALCULATION OF KAPPA:

P_o = sum of diagonal entries/total of all entries = 98/120 = 0.82

$P_c = \Sigma(P_1 \times P_2) = (0.217 \times 0.217) + (0.25 \times 0.25) + (0.15 \times 0.225) + (0.15 \times 0.092) = 0.18$

$$Kappa = \frac{P_o - P_c}{1 - P_c} = \frac{(0.82 - 0.18)}{(1 - 0.18)} = 0.78$$

B. An example of Kappa values using the same sampling method by more than one observer simultaneously observing the same animal.

Kappa Values	Sample Method/Interval		Group
0.78 - 0.96	instantaneous	/15 sec.	Jerry, Tana, Dale
0.94 - 1.00	one zero	/10 sec.	Diane, Mark, Barb
0.79 - 0.95	instantaneous	/10 sec.	Ed, Jane, John
0.15 - 0.90	predom. activity/	5 sec.	Andy, Carol, Doug
0.61 - 0.83	instantaneous	/15 sec.	Kirk, Theresa, Ken

C. An example of some hypothetical data gathered by 2 different observers watching the same animal, illustrating the calculation of Kendall's coefficient of concordance, W (from Sackett, 1978).

OBSERVER	BEHAVIOR CODES[a]					CALCULATION OF W
	A	B	C	D	E	

$$W = \frac{\Sigma(R_j - \bar{R})^2}{1/12k^2(N^3 - N)} = \Sigma(R_j - \bar{R})^2 =$$

1	2	1	4	3	5	
2	2	1	3	4	5	
R_j	4	2	7	7	10	

$$(4 - 6)^2 + (2 - 6)^2 + (7 - 6)^2 +$$

$\bar{R} = \Sigma R_j/N = 30/5 = 6$

$$(7 - 6)^2 + (10 - 6)^2 = 38$$

N = number of behavior codes
= 5
k = number of observers = 2

$$W = \frac{38}{(1/12 \times 4 \times [125 - 5])} = \frac{38}{40} =$$
$$= 0.95$$

[a]Scores are rank order of each behavior for each observer

D. An example of values for Kendall's coefficient of concordance, W, for ordinally ranked data.

W Score	Comparison Made	Sampling Method	Group
0.92	same method/ different observers	instantaneous	Jerry, Tana, Dale
0.83	different methods/ same animal	classical one-zero	Diane, Mark, Jo
0.75	different methods/ same animal	predominant activity	Marla, Ken, Kirk

4. ORIENTATION: I. THERMOKINESIS IN <u>TRIBOLIUM</u> BEETLES. R.D. AKRE

1. Construction of the Constant Temperature Apparatus

The test apparatus is constructed of 20 cm diameter clear plastic pipe (5 mm wall) (Fig. 4-1 in exercise). Place the inlet and outlet tubes as shown for good circulation and a minimum of problems in maintaining temperature differentials. The inlet tube should always be the single tube directly opposite the divider; outlets, the 2 tubes next to the divider. The flow of water coming in never exceeds the rate at which it can leave the box, and the box never ruptures.

Tops (2.5 mm) and bottoms (5 mm) of the test apparatus are Plexiglas or similar clear plastic sheets. The divider is constructed of two 2 mm clear sheets glued together and then glued into the box (for strength). Glue is industrial strength epoxy resin; bathrub caulk or aquarium sealer will also work well.

A 1.5 cm piece of the same pipe is cemented to the top of the apparatus to hold the beetles on the top surface, and to hold glass plates above the beetles to trace their routes. Surfaces on which the beetles walk must be roughened with sandpaper to prevent slipping.

Unfortunately, these test arenas are not available commercially. They can be manufactured by most shops (University Physical Plants) for $60 to $150 each. While this seems expensive, well-constructed arenas should last for 10 years or more.

2. Laboratory Set-up

The apparatus is set up as shown in Fig. 4-1 of the exercise. As the apparatus fills with water, air bubbles frequently form on the top surface of the box. Tip the box while holding the bottom outlet tube closed, and the water flow will rapidly expel the bubbles.

3. Exercise Length

Depending upon the length of the laboratory period, the activities that follow may be divided among teams in different ways. You may wish to set up the water bath equipment as 2 stations, one with a 20°C temperature differential and the other with a 30°C differential, and alternate use of the stations so that all teams perform both experiments. Alternatively, you may wish to assign half the teams to each differential and pool class results. (Similar decisions might also be made concerning use of the 2 species.) If the instructor sets up the lab exercise and achieves the desired temperatures in the test arenas before the class starts, the exercise takes about 2.5 hr. This assumes that duties are divided among different student groups, and analyses of results are concurrent with the experiment.

---

5. ORIENTATION: II. OPTOMOTOR RESPONSES OF DRAGONFLY NAIADS. R.D. AKRE

A. Prelaboratory Preparations

If dragonfly naiads are collected in the spring from cold water ponds, they should be placed in an aquarium with pond water several days prior to the laboratory, and allowed to warm to room temperature gradually. With a minimum of care, the naiads will remain alive for several months or more.

B. Instructions for Optomotor Apparatus

The treadmill optomotor apparatus is constructed as shown in Fig. 5-1 of the student exercise. Treadmills also can be constructed from old beer signs (the ones with rotating scenery) or rock polishing tumblers. Merely place a striped pattern between the rotating shafts to make a treadmill. Attach a variable speed motor to the shaft (detach motor on beer sign or tumbler), and connect to a rheostat. This will permit some control of the speed at which the treadmill revolves. A gear reducer permits even more variation in speed.

The optomotor apparatus constructed from the variable speed drill, speed reducer, and several plastic dishes is much less expensive to make (most items are available in the school shop). The drill and dishes are supported in the vertical position by ring stands and clamps.

C. Preparation of Artificial Prey Items

Glue small pieces of plastic to wire or wooden handles. Paint various patterns of stripes on the pieces of plastic. Other prey items are easily constructed; for legs, use rubber bands, and for bodies, pieces of cork or small rubber erasers. Fishing lures with their hooks removed could also be used. Because naiads appear to perceive prey objects only as indistinct shapes, all manner of artificial prey will evoke a strike response.

6. PATTERNS OF SPATIAL DISTRIBUTION IN FIELD CRICKETS. W.H. CADE

1. This exercise could be done with a variety of acoustical insects. Grasshoppers, katydids, and crickets are suitable. Grasshoppers and katydids which occur on long stems of grass are sometimes more easily disturbed than are ground dwelling crickets, making the crickets more suitable. Members of the genus Gryllus (field crickets) are widely distributed in North America and are very common in the early fall. They are all ground dwelling, black to brown in color, and measure 14-30 mm. Gryllus inhabit forested areas and almost any grassy area. The ideal location is a nearby park where vegetaion is low and the flags are easily located once placed. Crickets are easy to approach as well when the vegetation is at a minimum. In picking a species, the important criterion is that there be at least 20-40 individuals chirping in the area to be sampled. A suitable area must be located before lab and the 100 m^2 plot are laid out.

2. Crickets must be located when most active. A few species show a great increase in calling just before dawn, but this is probably not the case with most Gryllus. With the northeastern Gryllus, the number of males chirping is very constant throughout a warm night. On cold nights, approximately the same number of males call in the day. Although it is possible to have students determine the activity schedule before the spacing lab, somewhat casual observations for the few days preceeding the lab should yield sufficient information.

3. Depending on the number of students, it may be desirable to break the laboratory exercise into seperate parts. All students should have experience at listening to and locating crickets, to allow direct observation of cricket behavior. Since some students will be better at this than others, the actual marking of calling males might be left to this group. Establishing the grid and mapping locations might be a separate exercise, and taking the nearest neighbor measurements another. Students would then pool their data.

4. Determining the precise location of calling crickets takes much time and patience. Beginning students should be expected to pinpoint the location of crickets within about 0.5 m. There invariably will be errors in these measurements.

5. The exercise calls for the grid to be established using flags (tape tied to stakes is best) and cricket positions to be marked with tape tied to vegetation. Where vegetation is low, the grid can also be marked using 2 different colors of tape tied to the vegetation.

---

7. OBJECT ORIENTATION IN MEALWORM BEETLES. R. JANDER

---

1. Maintaining and Using Mealworm Beetles

These experiments can be considered almost foolproof if the following simple and obvious prescriptions are followed.

a. For good results, maintain a well cared-for culture of mealworm beetles with a mixture of all instars to ensure the availability of prime beetles at the proper time. Supplement the staple food of bran with fresh plant food such as lettuce or slices of potatoes. It is an encouraging sign if uncovered beetles quickly run for cover. If you rely on beetles newly shipped in from some biological supply company, you are risking trouble. Everyone can afford the little time necessary to properly maintain such a culture.

b. Stress to your students that, as in most experiments with the behavior of mealworm beetles, these experiments usually only work in dim light and with healthy active beetles that have not been handled roughly.

c. For additional pointers, see Exercise No. 15, Communication in Mealworm Beetles.

2. Prelaboratory Preparations

In order to water-deprive the beetles, take all moist food away 1 week prior to experimentation. Select active and undamaged beetles, and temporarily keep them in a container with places to hide, such as between sheets of filter paper or cloth. This precaution reduces mutual agitation between beetles and the possibility of damage to them.

3. Construction of Equipment

Pieces of cardboard should be somewhat roughened on their surface to provide good hold for the beetles' claws, since mealworm beetles lack adhesive pads on their tarsi.

The elevated T-mazes and L-T mazes are made from wooden dowels of 1/4 or 3/8 inch diameter. Cut sections are glued together and supported by needles, which in turn are inserted into a base plate (Fig. 7-2a in the exercise). A minimal set of mazes is illustrated (Fig. 7-2b,c); the long first section of all these mazes is 15 cm. The T-bars at the terminal ends are 6 cm long, and the section between the T-bar and the L-bend is 3 cm long. There is no limit, however, to the inventiveness of students or their instructor.

4. Class Discussion

   The series of experiments on ranging are well suited to lead
into a discussion of the proper control of random errors and
systematic errors in behavioral experiments. They tie in well with a
later exercise (No. 12) on randomness and determinedness of local
search in honey bees foraging on artificial flowers, and another on
stimulation of this phenomenon by stochastic models (Exercise No.
32).

---

8. THE LOCOMOTION OF GRASSHOPPERS. R. FRANKLIN

   Grasshoppers should be collected or ordered and placed in cages
about a week prior to the planned lab. Put them in roomy cages with
a light bulb inside to provide a light/dark cycle (12 hr.:12 hr.) and
warmth. They can be fed lettuce. The interior of the cage should be
moistened by spraying it with water from an atomizer once a day.
Ideally your animals should be used 1 to 2 weeks after their final
moult. Older animals tend not to behave as well as often in
experimental situations.
   The experiments on rotation in walking require a large,
vertically oriented optomotor drum approximately 60 cm in diameter
with a pattern (see below) 60-80 cm in height. The inside center of
the drum should be stationary. This can be accomplished by using the
type of bearing used for "lazy susan" turntables, with the center
platform supported through the center of the bearing. A variable
speed motor is needed to rotate the drum at different speeds (see
Exercise 5, Figs. 5-1 and 5-2). Using an elastic band from a pulley
on the motor shaft and wrapped around the base of the drum usually
works well.
   The black and white stripe patterns are designed to represent an
abstraction of plant stems in the case of the swimming experiments,
and as a simple orderly pattern in the optomotor drum. Different
species of grasshoppers probably prefer different stripe widths, but
in general a width of 2 to 2.5 cm made as long as your pattern
requires will work quite well. Spraying the patterns with an acrylic
preservative helps them last longer.
   Dental wax may be obtained from Lactona Corp., Academy and Red
Lion Roads, Philadelphia PA 19114.

9.   OPTIMAL FORAGING STRATEGY: I.   FLOWER CHOICE BY BUMBLE BEES.
     B. HEINRICH

Capillary tubes and a refractometer can be purchased from most
scientific supply companies. (A direct distributor of the Bellingham
and Stanley Pocket Refractometer, with range of 0-50% sugar, is Epic
Inc. 150 Nassau Street, New York, NY 10038, Tel. 212-349-2470).
     To create an artificial flower "field" may require several
hours. Be sure to put flowers into water immediately upon cutting in
order to prevent wilting. All of the stems should be of similar
height.

10. OPTIMAL FORAGING STRATEGY: II.   FORAGING MOVEMENTS OF BUMBLE
     BEES. B. HEINRICH

This experiment will require some forethought and advanced
planning. A week to 10 days before the laboratory, find a lawn with
white clover in it, and arrange to leave it unmowed. In a week or
less it should be covered with a carpet of flowers, and be actively
visited by many bees.
     One to 3 days before class, select 2 study plots, 3 m^2. String
a series of grid lines, using white string and appropriate stakes, 15
cm apart at right angles to each other. Each plot should measure 20
lines to a side. Cover one of the plots with fine mesh screening;
hold down the sides of the screen with stones or sticks so that no
bees can get in under it. Leave the screen on for 1-3 days so that
nectar can accumulate in the flowers. If the class is large, you may
wish to replicate these plots in order to reduce congestion during
the field exercise.

11. FORAGING BEHAVIOR OF SUBTERRANEAN TERMITES.   W.L. NUTTING

This exercise obviously depends upon finding an area where termites are fairly common. While they occur throughout the United States, except Alaska, subterranean termites are only abundant in parts of the southeast and locally common elsewhere (Weesner, 1965, see references at end of the exercise). They are active the year around where there is little or no freezing, and during a progressively shorter season with increasing latitude and elevation. There is usually someone knowledgeable about them in most university communities. If they are an economic problem, a local pest control operator might be willing to provide help in locating an infested area. The instructor should scout promising areas in advance of the exercise. A few rolls of paper could be set out to assist in identifying suitable sites and testing the potential attractiveness of particular papers.

This method has been very successful in arid and semi-arid areas of the western and southwestern U.S., Australia, and Egypt (ca. < 300 mm precipitation) (LaFage, Nutting and Haverty, 1973; Johnson and Whitford, 1973). Trials in areas of higher rainfall (ca. > 1000 mm) indicate that the rolls are not attractive when excessively wet. This problem can be minimized or avoided by wrapping or covering the rolls in various ways.

The best way to locate termites is to walk slowly through an area, turning over stones and every bit (above a few cm in length) of woody litter, cardboard, and dung. Gnathamitermes and some of the Amitermes build conspicuous tubes and broader shelters over dead plant material, sticks, logs, tree trunks, fence posts, etc. If any termites are present, a half-hour's search should turn up foraging groups or vacant galleries. Such a search should be made during warm weather, preferably when the soil is damp.

While this exercise extends over many weeks, it does not require a great deal of time for the weekly samples, and therefore can be carried on concurrently with other laboratory work. Since colonies apparently live for decades, a good site can be used for years. You may wish to consider establishing your plot in cooperation with another class (in ecology, for example) for use concurrently or in a succeeding semester.

Security is a matter which will have to be worked out on an individual basis. Consider private or school property, farms, local preserves, or use informative signs. The only vandalism experienced on a U.S. Forest Service experimental range in southern Arizona was occasional "displacement" of a few rolls by playful coyotes. The wire anchors will prevent their being carried away by animals or wind.

Instead of the square grid, you might wish to set the rolls out on a transect: Through gradients or abrupt changes of vegetation,

soil, slope, exposure, etc., or even around a house or building known
to be infested with termites. The transect might be more effective
in the form of 2 rows, a meter apart, but with the rolls staggered as
in a continuing "W".

---

12. FORAGING ORIENTATION OF HONEY BEES. R. JANDER

---

1. Construction of Flower Models

A simple way of constructing the flower models is the following
(see Fig. 12-1 in the exercise). Cut a broomstick into sections, 6
cm long. On each end glue circular discs of 6 cm diameter. A
convenient material for the discs is 32 mm (1/8 inch) masonite. Into
the top center of the model drill a vertical hole which is 2 cm deep
and has a conical bottom. The diameter of the hole is 6 mm. Its
inside should be waterproofed with wax or some other equivalent
lining. Finally, the flower models are colored with a waterproof
paint. In choosing a pattern there is much room for imagination, but
all flowers should be the same. The base, for instance, might be
green and the top, blue with radiating white lines. Fully white
flowers are not recommended, however, because they are apparently
less well remembered (Menzel R., J. Erber and T. Masuhr, 1974).
Learning and memory in the honey bee. pp. 195-217 in Experimental
Analysis of Insect Behavior, L. Barton-Browne (ed.), Springer-Verlag,
New York).

2. Prelaboratory Preparations

In addition to constructing the flower models and preparing the
sugar or honey water solutions, you must train the bees to the
location of testing several days in advance of the foraging
orientation activities, unless you wish the students to participate
in this phase of the activity.

Because honey bees have a strong tendency to orient upwind, you
will want to either choose a testing location which is very
sheltered, or record natural wind direction frequently during the
experiment.

13. HOST ACCEPTANCE BY APHIDS: EFFECT OF SINIGRIN. L.R. NAULT

1. Obtaining and Rearing Aphids

If someone at your college or university is not culturing the aphids required for this exercise, they can easily be obtained from the field. During mid- or late summer, turnip and cabbage aphids may be collected from cole crops in non-sprayed vegetable gardens. Turnip aphids are more common on radish and turnip, whereas cabbage aphids are frequent late summer and early fall inhabitants of cabbage, broccoli, and cauliflower. The turnip aphid is a dark olive-green, whereas the cabbage aphid is light green and covered with a grey or white waxy "bloom" which gives heavily infested plants a whitish appearance. The third aphid species found on cole crops is the green peach aphid, a light green aphid with no waxy coating.

Pea aphids are commonly found on peas in the vegetable garden. They are the only large, light green aphid on this crop. The species can also be collected with a sweep net from alfalfa or red clover fields or from wild sweet clover.

Rear turnip and cabbage aphids on turnips or radish, and pea aphids on broadbean (see following instructions for planting). Remember these are parthenogenetic insects. They will produce 4-10 progeny per day and can easily "overcome" their hosts. Provide new hosts at least once a week, and be certain to water host plants frequently. Check Nault (1969, see references at end of the exercise) for other rearing suggestions. Build up your colonies 2 weeks prior to the exercise. Note that for section C alone, each student will need 50 adult pea aphids for the no-choice test.

2. Prelaboratory Preparations

Be certain to plant turnip or radish 4-6 weeks in advance of the exercise. These are small-seeded vegetables, and it takes a while for leaves of sufficient size for use to develop. The large-seeded broad bean needs to be planted 3-4 weeks prior to use. Potatoes can be started from tuber 4-6 weeks in advance.

Sinigrin is available from United States Biochemical Corporation, P.O. Box 22400, Cleveland, OH 44122. One gram dissolved in 100 ml of distilled water will give you a 1% or 10^4 ppm stock solution. Figures 13-1 and 13-2 in the exercise illustrate typical experimental design and results.

14. BLOW FLY FEEDING BEHAVIOR.   J.G. STOFFOLANO, JR.

1. Sarcophaga bullata can be substituted readily for Phormia regina, and are easily obtained from biological supply companies. Instructions for maintaining them come with the purchase of the flies. One should order the flies at least 3 wks. before intended use, and should specify the need for fresh pupae that have not been refrigerated.

2.   Applicator sticks can be obtained from most biological supply houses. Tackiwax is purchased from Central Scientific Company. The stick holders are made from any scrap wood at least 2.5 cm in thickness and 7-10 cm in width. The length should be about 25 cm, thus permitting holes to be drilled at intervals of 2.5 cm. The holes should be drilled just large enough and deep enough to permit the sticks to be inserted.

3.   Prior to testing a fly on the series of sugar solution, touch the tarsi to distilled water; if proboscis extension occurs, permit the fly to drink to satiation. Do not push the fly's legs into the solution, but just touch the tips. It may be easier to observe proboscis extension if you keep the fly and solutions at eye level. Also, do not let the tip of the proboscis touch the sugar solutions.

4.   It is a very simple procedure to anesthetize insects with CO_2. This is done by permitting the gas to enter the closed container housing the flies. Try not to use a container over 1 liter, otherwise it takes more gas to "knock them down." If the container has a screened top, cover it with your hand. Do not worry about killing the flies, since they can survive long exposures to CO_2. Once the flies are immobilized, transfer the container to the crushed ice. In order to put the flies on the sticks, do not remove them en masse from the ice. Remove one individual at a time to prevent them from becoming active.

5.   Section D, the removal of sources of negative feedback governing feeding, may be too difficult for most students. Flies can be made hyperphagic, however, using the techniques described by Nelson (1972). Dethier (1976) describes how to perform the transection of the recurrent and ventral nerves. (Reference citations given in exercise.)

15. COMMUNICATION IN MEALWORM BEETLES.   R.L. RUTOWSKI

1. Prelaboratory Preparations

One week or more before the laboratory session, the adult mealworms should be obtained and separated according to sex. Adults may be sexed as indicated in Fig. 15-1 of the exercise. Pupae may also be sexed with practice under a dissecting microscope. Male pupae have a small depression on the ventral surface of both the ultimate and penultimate abdominal segments. Female pupae have a depression only on the ventral surface of the terminal abdominal segment. Female pupae also have 2 small knobs or tubercles anterior to this depression that are not found on the male. Females may be housed in a group. Males are best housed individually in a petri dish with a small quantity of bran, a cube of potato (for moisture), and a piece of paper towel (for cover and footing). This procedure increases the willingness of males and females to engage in sexual interactions.

While isolating these males and females, kill 5 to 10 individuals of each sex with a killing jar. Let them dry out in the air at room temperature during this procedure, because once they are dry you will not be able to sex them.

The alcohol extract of female bodies should be prepared the day before it is to be used in the laboratory. Instructions for preparing the extract are in the exercise.

At the same time this extract is being prepared, place individuals of both sexes in a freezer and store them until needed as fresh-killed animals for the experiments. The goal is to have these beetles immobile during the presentation to males; if freezing does not kill them, use a killing jar to stun or kill them for the experiments.

2. Hints for the Instructor

After observing the behavior it is important to establish a bioassay procedure that all of the groups in a given lab period will use. As a guideline, experimental stimuli should only be tested on males whose sexual interest in females has been documented by observing their response to live females. Males should not be used if they do not court and attempt to mount females within 5 min after a female is introduced into their cage. All experimental stimuli are best left with each male for 5 min or until the male mounts the stimulus and extends his genitalia in an attempt to couple with it. It is recommended that the students work in pairs that are responsible for collecting data from at least 5 males on each of the stimuli presented.

A typical series of experiments and their results are shown in the table below. As can be seen, males still differentiate between males and females even when the stimulus animals are fresh-killed. However, males cannot discriminate among animals that have been dead for some time. In particular, females lose their attractiveness, suggesting that a critical chemical signal has been lost. This conclusion is further supported by the responses to the glass rods dipped in the alcohol extract. An additional stimulus (not shown on the table) that yields interesting results is female bodies that have been briefly washed twice in ethanol and then dried a few minutes before presentation to males. These females have generally lost their attractiveness to males.

Sample data set showing male mealworm beetles' responses to various stimuli.*

STIMULUS	RESPONSE	NO RESPONSE
Female - fresh killed by freezing	26	4
Male - fresh killed by freezing	0	30
Female - dead and air dried for 1 week	0	30
Glass rod dipped in clean alcohol	0	30
Glass rod dipped in female extract	24	6

*Each of 6 groups of 2 to 3 students collected data on 5 males.

16. CHEMICAL RECRUITMENT IN THE FIRE ANT. J. R. MATTHEWS

A. Choice of Ant Species

Solenopsis invicta, the imported fire ant, of economic importance in the southeastern U. S., is the preferred species and can be maintained in the laboratory for long periods after collection. If Solenopsis are unavailable, several other ant genera such as Pheidole, Crematogaster, Lasius, Acanthomyops, Formica, Monomorium and Iridomyrmex are potentially satisfactory. Solenopsis molesta, a minute yellowish ant commonly found under stones in open fields in the northern, eastern, and western states is also suitable.

Helpful keys to subfamily identification can be found in Snelling (Systematics of Social Hymenoptera, in Social Insects, vol. 2, pp. 369-453, H. R. Hermann, ed., Academic Press, New York, 1981) which also includes a list of genera included in each of the 11 subfamilies. Older but still useful keys to generic identification can be found in Creighton (Ants of North America. Bull. Mus. Comp. Zool., 104:1-585, 1950) and Ross et al. (A synopsis of common and economic Illinois ants, with keys to the genera [Hym., Formicidae], Ill. Nat. Hist. Survey, Biol. Notes no. 71, 22 pp., 1971). In general, species with a large colony size have had a tendency to evolve odor trail communication. However, many genera and species of ants do not lay odor trails, particularly among the more primitive groups.

If adapting this exercise to other ant species, advance experimentation and modifications may become crucial to success. For an entree into the literature on ant behavior, the books by Sudd (An Introduction to the Behavior of Ants, St. Martins Press, New York, 1967) and Wheeler (Ants. Their Structure, Development and Behavior, Columbia University Press, New York, 1910) are useful.

B. Obtaining Fire Ant Colonies

Necessary Equipment: Shovel, trowel, or spade; talcum powder; large containers such as buckets, trash cans, or steel washtubs. Generously coat the inner sides of the container(s) with the talcum powder; this treatment will make the sides too slick for the ants, preventing them from crawling out.

CAUTION: Recall that the fire ant gets its name because of the vicious sting which it possesses. While individual stings can be painful, normally they are not serious; however, a great many stings could result in considerable discomfort. While collecting fire ants, wear boots and tuck your pants inside.

Fire ants make large, very conspicuous, mounded nests easily located along roadsides, waste areas, old fields, pastures, and towns. They now range throughout southeastern U.S. from Texas to North Carolina. The following techniques for collecting their colonies can be applied with only slight modification to almost any soil-nesting ant species.

Locate 3 or 4 colonies, and make a test excavation in one to determine where, and how deep, the ants and their brood are located. In general, the part of the mound having the most visible activity on the outside is the region beneath which the ants and brood will be most concentrated. Usually this is on the sunny side of the mound.

After determining the colony to be collected, WORK SWIFTLY. Attempt to obtain the bulk of the nest in 2 or 3 shovel loads. (Quickly brush off any ants adhering to you before they sting!) Dump the shovels full of dirt, mound, and ants directly into the container. Collect as much of the brood as possible, because the queen will usually reside in the general area of the brood. Nests

collected without the queen will eventually die out. However, for
short term use, a queenless colony will be entirely satisfactory.

Carry colonies back to the laboratory. In 1-2 days, the ants
will settle down and reconstruct galleries and nest. Periodically
dust more talcum powder on the inner sides of the container as
needed. If Fluon® is available (see Aids, Exercise 3), a coating of
it provides protection of longer duration than the talcum powder.
Colonies should be provided honey and fed mealworms, crickets and
peanut butter. The soil should be watered lightly every few days.
A simple method of culturing fire ants using disposable petri dishes
is described by Bishop et al. (J. Georgia Entomol. Soc. 15:300-304,
1980). To separate ants from soil use a flooding technique in which
water is slowly added to the nest container. This drives the ants to
the surface as the container fills, eventually causing them to ball
up in a floating mass which can be scooped up and transferred.

C. Additional Sources of Information

A minicourse "Communication Through Odors" which includes an
investigation of trail laying and recruitment in ants has been
published in Investigating Behavior prepared by the Biological
Sciences Curriculum Study, Inc. and published in 1976 by W. B.
Saunders Co., Philadelphia. A separate Instructors Manual is also
available.

A super-8 film loop "Chemical Communication" shows some of the
techniques used in this exercise. It is available from Hubbard
Scientific Company, P. O. Box 105, Northbrook, Ill. 60062. (catalog
#9729)

---

17. EXPLORATORY AND RECRUITMENT TRAIL MARKING BY EASTERN TENT
CATERPILLARS. T.D. FITZGERALD

---

1. Establishing Colonies

A satisfactory artificial diet for the eastern tent caterpillar
has not been developed to date, and the caterpillars must therefore
be fed fresh cherry or apple leaves. Black cherry (Prunus serotina),
a preferred host, is found throughout the range of the caterpillar.
Use only the younger leaves, preferably those that are not fully
expanded, when feeding the first instar larvae. Leaves more than 3-4
wks old may be rejected or eaten reluctantly even by the larger
caterpillars. If the study is conducted in the spring, leaves will
be readily available. Branches collected in late winter and placed
in water will sprout and provide enough leaves to bring colonies
through the first 3 or 4 of the 6 larval instars.

Young colonies or overwintered egg masses can be collected in the spring and brought into the laboratory. If eggs are collected earlier, they must be refrigerated before use. Eggs require a minimum of 15 wks storage at 2°C before they will hatch (Bucher, 1959), but should not be subjected to cold treatment until 3-4 wks after oviposition to allow for completion of embryogenesis. Eggs collected in late winter will require proportionately less cold treatment. Eggs collected in the summer and stored at 2°C will remain viable for 6 months or more. During storage, eggs must be kept from dessication. A saturated solution of potassium chloride will provide an adequate R.H. of about 85% at 2°C.

Following cold treatment, eggs can be hatched at room temperature in petri dishes. If adequately cold treated, the larvae will eclose within several days to a week. Leaves can be added to these dishes and the larvae maintained in the dishes for the first 3 to 4 instars. This insect is adapted to a rigorous spring climate, and petri dish colonies or tents can be stored at 5°C for a week or more with no apparent harm to the caterpillars.

When leaves are not available, the trail-following behavior of individual larvae can be studied by using newly eclosed, unfed first instars. Their small size, however, makes them more difficult to handle than second or third instars which are otherwise preferable. Silk can also be collected during the spring and extracted or stored in a freezer until needed. Eggs collected within a month of oviposition and stored at 2°C as indicated will provide larvae for the fall semester (October or November).

2. Field Studies vs. Laboratory Studies

If natural populations are nearby, field studies of trail-following can easily be conducted. To facilitate a field study of recruitment, the instructor can defoliate several small infested trees the day before the class is scheduled. Laboratory studies of whole colonies (Sections A and B) are best conducted in the spring when leaves are readily available. These studies require more time (and are easily elaborated upon), and students might be given colonies to rear and study independently over a 2-3 week period. Alternatively, the instructor may wish to establish a laboratory colony and demonstrate recruitment during the class session.

EDITORS' NOTE: Exercise 7 also utilizes maze-running, and contains a discussion of sources of systematic error which raises points relevant to this exercise as well.

AUTHORS' NOTE: A paper published since the original submission of this laboratory exercise contains information that would be useful in designing additional laboratory studies.

Fitzgerald, T. D. and J. S. Edgerly. 1982. Site of secretion of the trail marker of the eastern tent caterpillar. J. Chem. Ecol. 8: 31-39.

18. DOMINANCE IN A COCKROACH, NAUPHOETA. G.C. EICKWORT

The roach colony should be provided with food (laboratory chow) and water ad lib., and with paper towels for shelter. Best results are obtained when the colony is maintained on an artificial cycle of 12 hrs. light : 12 hrs. dark, with the cycle timed so the exercise will be conducted during hours 2 to 4 of the dark part of the cycle. Observations should be conducted under red light, in an otherwise darkened room.

About 2 months before the exercise is to be conducted, anesthetize the culture with carbon dioxide. Remove the largest nymphs and the teneral (soft, pale) adults. Sort these under the stereomicroscope by sex (Fig. 18-1 in the student exercise). Maintain males and females in separate culture tanks with food and water. Be especially careful that no males are in the female tank.

About 10 days before the exercise, place those roaches which have moulted into adults into petri dishes, 1 roach per dish, with food and water. (There should be 3 male roaches per student, and 1 female roach for every 2 students. If insufficient male roaches have moulted to adults, obtain additional adult males from the culture tank.)

Three days before the exercise, take half the male roaches and combine them so there are 3 per petri dish. Anesthetize them, and mark each male in a dish with a different mark of white enamel or acrylic paint on the pronotum. These animals will form the established dominance hierarchy used in part A. The other males are kept in their separate dishes, but are also marked so that there are equal numbers of each of the 3 marks (1, 2, or 3 dots on the pronotum will distinguish the roaches under red light). These will be used in part C. Leave the females unmarked.

Food and water should be removed from the petri dishes just prior to the exercise. Empty petri dishes should be available for part C at the beginning of the exercise. As each pair of students finishes with parts A and B, 3 differently marked, isolated males should be removed from their dishes and placed together in an empty dish. These dishes should be marked with a wax pencil to distinguish them from those containing the established hierarchies. Female roaches that copulate in part D should be returned to the culture tank and not reused in a later lab section.

In large classes, the number of male roaches and petri dishes may be limiting. If there are multiple laboratory sections, the established male hierarchies can be reused for each section, but females must be virgin and consequently cannot be reused once they have mated. The newly established male groups in part C are best not reused, but if they are immediately returned to separate petri dishes after an exercise, they can be reused in different combinations in subsequent sections.

19. CRICKET PHONOTAXIS: EXPERIMENTAL ANALYSIS OF AN ACOUSTIC
RESPONSE. G.K. MORRIS AND P.D. BELL

Many variants of this exercise are possible. Calling songs of
acoustic Orthoptera are species-specific, and mature virgin females
will discriminate the songs of conspecifics from those of other
species (Morris et al., 1975). In season, many cricket species can
be collected as late instars in the field. They should be kept
sexually isolated in suitable cages with dry dog food and apple
slices. About 2 weeks after their molt to adult, females should show
strong phontactic responsiveness to the recorded calling song of
their conspecific male. Isolate later instar nymphs and use them in
the experiment no sooner than 1 week after their molt to adult.
Non-virgin adult females will be receptive if kept isolated from
males for at least 2 weeks. Virginity in adults that are mature is a
guarantee of receptivity, but freshly molted individuals may require
some days to become sexually active.

Tree cricket, Oecanthus spp., (Fig. 19-1 in the exercise), and
common black field crickets, Gryllus spp., can be taken from the
field even as adults. After sexual isolation for a week or so, they
should show receptivity even if they have mated previously. Late
instar nymphs of Oecanthus may be obtained by beating roadside or
old-field vegetation with a net. Tree crickets can be particularly
useful, as much is known of pair-forming responses in these insects
(Walker, 1957). Oecanthus nigricornis can be collected out of season
by searching for egg scars on raspberry canes. Specimens can be
raised in the laboratory from field-collected eggs (Bell, 1979).

Tape recorders can be cassette or reel-to-reel. It is important
that their frequency response extend up to at least 10 kHz to
properly store and reproduce the frequencies of most crickets. A
good way to test the efficacy of any playback system is to observe
the speaker output on an oscilloscpe. First the output of a signal
generator (or the insect's song for that matter) is viewed on the
oscilloscope and recorded. It is then played back through the arena
system while a microphone monitors the broadcast sound. The
emergence of a signal which is a reasonable facsimile of the one
originally recorded gives assurance that the equipment is working
properly. Oscilloscope displays can also be used to set sound
broadcast levels which are consistent between replicates or
subsequent experiments by positioning the speaker and microphone at
a fixed measurement distance and aspect. It might be advisable to
have master tapes, from which duplicates are made for student use.

20. DEFENSE IN BEETLES.   J. R. MATTHEWS

1. Obtaining Beetles

Mealworms are often available from live bait dealers, and are easily maintained for long periods in the laboratory. They may also be obtained from biological supply houses, which usually enclose a brochure describing culture methods. We maintain them in large plastic shoe boxes, filled to a depth of 3-4 cm with rolled oats. Occasional supplements of potato or apple slices provide additional nutrients and moisture. (See also Aids to the Instructor for Exercises 7 and 15.)

Dermestid beetles of various species are also available from biological supply companies or from sources listed in Arthropod Species in Culture (listed in Appendix 3). Various species may be readily collected in warm seasons from old nests of mud dauber wasps where they scavenge the dried prey remains in aborted cells. (NOTE: In the laboratory, dermestids should be kept in tightly sealed containers, or they may infest insect collections and other dry organic matter, causing serious damage.) Dermestids will reproduce satisfactorily over several months in a tightly closed container if periodically provided with a supply of dried insects. We have also used old paper wasp nest combs containing unemerged pupal cells as a food source.

Species of ladybird beetles (Coccinellidae) are common almost everywhere. Large numbers of adults may sometimes be collected from late fall to spring in protected areas under boards, rocks, etc., where they hibernate in dense aggregations. However, locating such aggregations may be difficult. Some instructors have successfully advertised in local papers. Commercial sources of ladybugs include advertisements appearing in various nursery and seed catalogs (e.g. Gurney's) and gardening magazines. Also, certain species are regularly cultured in some entomology laboratories. Sources for these may be found in Arthropod Species in Culture. Adult coccinellids are voracious predators on aphids, and difficult to maintain over long periods. Either plan to have beetles shipped just prior to intended use date, or store them in the refrigerator until use, which mimics hibernation conditions for them. In warmer seasons, numerous individuals may be collected by sweeping nets in fields where aphids are abundant. Occasionally one finds shrubs so heavily infested with aphids that a sizeable resident population of coccinellids has built up and can be hand picked.

2. About Chickens

We have successfully arranged to "borrow" 5-10 hens from a friendly local farmer, picking them up early in the morning on the day of the class and returning them later the same day. Assure the farmer that none of the experimental foods to be offered will adversely affect the chickens' health. Experience to date has been that the various bantam breeds are the more "interesting" performers. Chickens can be easily transported in covered cardboard boxes. They perform very well if allowed to become slightly hungry by preventing access to food for 2-3 hrs prior to use. Suitable individual lab cages can be made by inverting plastic laundry baskets on lab tables which have first been covered with old newspaper. Water in a finger bowl or other stable dish should be continually available. Individual chickens often vary considerably in their pecking and eating behaviors.

3. About Chemicals

Salicylaldehyde is available from either Pfaltz and Bauer Co., 375 Fairfield Ave., Stamford, CT 06902 or Aldrich Chemical Co., 940 W. St. Paul Ave., Milwaukee, WI 53233. The dilutions should be prepared in advance of the laboratory session, and bottles clearly marked. Suggested concentrations are 50%, 10%, 5%, 1%, and 0.5% by volume. Be sure to have sufficient numbers of micro-pipettes available, and caution students to keep track of which are used for which solution, or contamination will occur.

The quinine dihydrochloride can be obtained from the above, and diluted.

4. Marking Mealworms

Experiment in advance with the markers or paints to be used to color the mealworms. Some colors of felt tip markers are not particularly distinctive on mealworms. Those which contrast strongly, such as blue or black, are preferable. Paints will produce even brighter colors.

5. General Comments

This exercise is purposely somewhat less rigorously structured and more open-ended than some others. Depending on how well it is organized in advance, the lab may require more than 3 hr to complete. Being opportunistic and flexible in approach is recommended. Encourage students to search for various beetles (and even other arthropods, e.g. terrestrial isopods) to contribute to the group used in Part A.

If time permits, have students prepare a water or glycerine temporary mount of the dermestid abdominal setae on a microscope slide. Alternatively, prepare such a slide in advance and have it set up as a demonstration (see reference to Nutting and Spangler, 1969).

If you have sufficient beetles and access to other potential predators (e.g. mice, gerbils, chameleons, toads, praying mantids, or salticid spiders), unusual and interesting observations on beetle defense may be made for comparison with those obtained in parts A and B. Other repellant chemicals may be substituted in parts C and D. For example, we have found that parabenzoquinone is very repellant applied to mealworms in a 16% solution. Ideas for other candidates for repellant compounds can be found in Blum (1981).

Much of the inspiration for the experiments in Part D came from Brower (1960). Other papers of particular relevance to the mimicry experiments include those of Sexton (1960), Schmidt (1960), Duncan and Sheppard (1965), and Morrell and Turner (1970). A good general reference on defensive behavior in animals is Edmunds (1974).

The film, "Do Animals Reason?", (14 min, color) produced by the National Geographic Society in 1975, includes a segment on starlings given choices between treated and untreated mealworms of different colors, and illustrates how rapidly avoidance learning can occur.

6. References (in addition to those included in Exercise 20)

Blum, M.S. 1981. Chemical Defense in Arthropods. Academic Press, New York.

Duncan, C.J., and P.M. Sheppard. 1965. Sensory discrimination and its role in the evolution of Batesian mimicry. Behaviour 24: 269-282.

Edmunds, M. 1974. Defence in Animals. Longman Inc., New York.

Morrell, G.M., and J.R.G. Turner. 1970. Experiments on Mimicry. I. The response of wild birds to artificial prey. Behaviour 36: 116-130.

Schmidt, R.S. 1960. Predator behaviour and the perfection of incipient mimetic resemblances. Behaviour 16: 149-158.

Sexton, O. J. 1960. Experimental studies of artificial Batesian mimics. Behaviour 15: 244-252.

21, 22. REPRODUCTIVE BEHAVIOR OF GIANT WATER BUGS. R.L. SMITH

Species representing 3 genera of giant water bugs can be found in the United States. Members of the genus Lethocerus, "electric light bugs," are very large insects frequently found at street lights during their summer dispersal flights. Lethocerus females lay their eggs on emergent vegetation (e.g. Typha) above the surface of the water and neither sex broods eggs. This is the ancestral method of reproduction, and the genus Lethocerus is believed to have given rise to the genera characterized by male brooding -- Belostoma and Abedus.

The 2 most common species of Abedus, A. herberti Hildago from Arizona, southern Utah, and southwestern New Mexico and A. indentatus (Haldeman) from California and western Nevada, are suitable for the outlined water bug exercises. (Other species of Abedus and members of the genus Belostoma are difficult to work with in the laboratory, but may provide interesting variations on the outlined exercises.)

Both preferred species of Abedus are stream dwellers, available in abundance from scores of localities within their respective ranges of distribution (see Menke, 1960). They can be collected easily with a large kitchen strainer or aquatic net in patches of watercress and emergent vegetation at the lateral margins of slow-flowing sections of streams. In the absence of aquatic macrophytes, bugs may be found by turning rocks in pools or by collecting at night with the aid of a gas lantern. The 2 species are allopatric and ecological replacements, one of the other. For purposes of these exercises their behavior patterns are virtually identical.

NOTE: Mixing the 2 species will produce ambiguous experimental results. Live specimens of the 2 are indistinguishable to the layman except by collection locality. It is best to use 1 species exclusively.

Abedus spp. are long-lived (> 15 months) and easily kept in the laboratory; however, immatures must be isolated from one another to prevent cannibalism, and adults should be isolated until paired for mating. One liter plastic cups (or quart plastic ice cream containers) are suitable. Each should be half-filled with distilled water and provided with a rough angular stone and some floating aquatic plants. Weekly, adult males should be fed 1 cricket each. Unless conspicuously gravid, females should be fed 2 crickets each week, in which case 1 cricket per week should be adequate. About 30 days are required for females to become heavily gravid after having laid eggs or undergone the definitive molt. Crickets may simply be dropped on the surface of the water where hungry bugs will instantly grab them.

Abedus spp. may be reared in the laboratory with modest effort (see Smith, 1974). Those living far from the southwest may wish to consider this option. When eggs hatch, the first instar nymphs should be quickly isolated in small plastic cups (such as 15 cc condiment cups) half-full of water and containing a small stone. Young nymphs may be fed vestigial-winged Drosophila, wing-clipped houseflies, mosquito larvae, or other prey of appropriate size. After the second molt, the third instar nymphs should be transferred to 1-liter containers and maintained in the same manner as adults. The hind legs of crickets should be removed before presentation to nymphs to lessen the risk of injury to the young bugs.

NOTE: Dirty water is the primary cause of mortality in rearing water bugs. A critical aspect of maintaining adults in the laboratory is that the water be changed 1 or 2 days after each feeding.

The molts of these insects are predictable to within 1 or 2 days. A few minutes prior to molting, the bug grabs its middle legs with the raptorial front legs while floating on the surface of the water (see Smith, 1974, 1975). Consequently, students should be able to observe molting, describe the process, and discuss how this molting behavioral pattern may have evolved.

My research on the bioacoustical behavior of water bugs has just begun, but it is already clear that the sounds are produced only in the context of courtship and mating. Males produce soft buzzing sounds while "display pumping" in the presence of a female. Females may occasionally answer a signaling male acoustically, but usually respond by attempting to oviposit on his back.

The best arrangement for hearing these sounds consists of a hydrophone (aquatic microphone) or a harmonica contact microphone waterproofed with a coating of silicon rubber. Hydrophones are available from Celesco Transducer Products, 7800 Deering Ave., P.O. Box 1457, Canoga Park, CA 91304. A harmonica contact microphone can be purchased at any music store. The hydrophone may be used to reveal acoustical communication in a variety of aquatic insects, both in the laboratory and in the field.

Signals from the microphone are fed to a 1 watt amplifier, and the amplified signal used to drive a headset (earphones). Feedback is a problem when a speaker is used. The microphone will respond best if an acoustically active male is sitting directly on it. This can be facilitated by using silicon rubber to glue the microphone to a stone.

Additional References

Cullen, M.J. 1969. The biology of giant water bugs (Hemiptera: Belostomatidae) in Trinidad. Proc. R. Entomol. Soc. Lond. (A)44: 123-137.

Lauck, D.R. and A.S. Menke. 1961. The higher classification of the Belostomatidae (Hemiptera). Ann. Entomol. Soc. Am. 54: 644-657.

Menke, A.S. 1960. A taxonomic study of the genus Abedus Stål. (Hemiptera:Belostomatidae). Univ. Calif. Publ. Entomol. 16: 393-440.

Smith, R.L. 1974. Life history of Abedus herberti in central Arizona (Hemiptera:Belostomatidae). Psyche 81: 273-283.

Smith, R.L. 1975. Surface molting behavior and its possible respiratory significance for a giant water bug Abedus herberti Hidalgo (Hemiptera:Belostomatidae). Pan-Pacific Entomol. 51: 259-267.

23. SEXUAL BEHAVIOR OF DAMSELFLIES.   J. Alcock and R. L. Rutowski

In order to help narrow the identification of your local damsel-flies, the following information is provided. (See also D. J. Borror and R. E. White, A Field Guide to the Insects, Houghton Mifflin). Damselflies resemble dragonflies but are generally smaller and much more slender, with a long thin abdomen; moreover, most damselflies hold their wings together above the body when they are perched whereas dragonflies always perch with their wings held out flat. There are three families of damselflies in North America. The species of the Calopterygidae are relatively large with colored wing markings. Males tend to be highly territorial along streams or rivers. The most widespread eastern species, Calopteryx maculata, has striking jet black wings and a beautiful iridescent green body (the male); the female is duller with dark grey wings and body. Members of the Coenagrionidae are generally small, delicate species, often with blue bodies and clear wings. They are associated with still waters and are usually non-territorial. The common genera are Enallagma, Argia and Ischnura. Damselflies of the family Lestidae look like miniature, thin-abdomened dragonflies because they hold their wings flat while perching. Some common species are greenish in color. They too inhabit ponds and swamps, rather than moving water. The common genus is Lestes. G. H. and J. C. Bick have published a bibliography by species of all papers on reproductive behavior in the damselflies through 1979 (Odonatologica 9: 5-18).

At the procedural level of organizing the laboratory we have found it productive to pose one problem at a time, then permit the students as individuals or in pairs to collect preliminary data for 30-60 minutes, and then meet again to discuss their interim results. At this time the class can be encouraged to formulate a standard procedure for data collecting that all class members will adhere to. Then the class is released to gather new data in the agreed-upon manner which is pooled and eventually distributed to the class members for analysis in their written reports.

24. COURTSHIP OF MELITTOBIA WASPS. R. W. MATTHEWS

1. Collecting Melittobia

Although not available commercially, Melittobia wasps can be easily field-collected at any time of year in most regions of the U.S. and throughout much of the world. Under highway bridges and on walls of buildings near water, one can usually find the conspicuous mud nests made by mud-daubing wasps. (Trypoxylon politum, the pipe-organ wasp, and Sceliphron caementarium, the black and yellow mud-dauber are the 2 common species in eastern U.S.) Most such nests, particularly where they are abundant, suffer high levels of natural mortality due to Melittobia infestation. Pry several such nests loose from the substrate with a knife. In the laboratory, break open the nests and obtain the cocoons. Gently open the cocoons; parasitism is signalled by the presence of a mass of tiny maggot-like larvae (see Fig. 24-1 in the exercise) instead of a single host larva or pupa. A single parasitized mud dauber cocoon will yield Melittobia sufficient to establish several cultures -- between 100 and 200 adults, of which over 95% will be the black, winged females. If host cocoons are collected any time from late fall through spring both the host and parasite will be in diapause condition as pre-defecating prepupae. In this condition they can be stored almost indefinitely in a refrigerator. When brought back to room temperature, development resumes, providing at least 2 months of exposure to cold has elapsed. Once defecation occurs (see Fig. 24-1 in the exercise) development to the adult requires an additional 1 month.

2. Culturing Melittobia

Transfer the contents of an infested host cocoon to gelatin capsules (size # 0, available from drug stores). Upon emergence, unattended adults will soon chew through their capsules. To prevent their escape, capsules may be kept in cotton- or foam rubber-stoppered shell vials. Adult females are relatively weak fliers and exhibit a strong positive phototaxis. If placed on a table top they will immediately begin to walk, hop, and fly in the direction of the brightest window or other light source. Take advantage of this behavior when handling and transferring wasps.

To reculture, shake a few adults onto a paper and quickly invert the small half of a gelatin capsule over them as they move toward the light. They will crawl upwards into the capsule and remain trapped there until needed. Often 2-4 adults can be captured in the same capsule half. In the other (larger) half of the capsule place a prepupa of Trypoxylon or a fresh pupa of the blow fly, Sarcophaga bullata (available from biological supply houses), then quickly and gently close the capsule halves together. (It is a good idea to first pop off one end of the fly pupal case to provide ready access for the _Melittobia_ and to ascertain that the host is fresh.) _Sarcophaga_ pupae store well in a refrigerator for several months and can be used as the need arises to reculture.

A thriving stock culture can be maintained by placing several adult females in a larger, tightly covered plastic container together with a number of _Sarcophaga_ pupae or host larvae or new pupae. Simply adding new fresh hosts at 3 to 5 week intervals will maintain a continuous supply of adults and immatures from which new cultures can be started to obtain known age adults.

3. Prelaboratory Preparations

Three to 4 weeks prior to the laboratory session, several _Melittobia_ cultures should be initiated. A few days prior to class, check the cultures to ensure that pupae are present. When the _Melittobia_ have reached the late pupal stage (recognizable by the change from whitish to black body pigmentation), the contents of the capsule should be shaken out onto an index card and viewed under low magnification of a dissecting microscope. Separate the sexes at this stage. Male pupae are amber or honey-colored and lack eyes; female pupae are black and have prominent reddish eyes. Use a camel's hair brush to tease pupae apart and transfer pupae to separate gelatin capsules. This ensures that virgins will be available for use in the exercise. Male pupae must be isolated, one to a capsule, because as adults they are extremely pugnacious and will fight and kill one another.

India ink can be used to write the dates and sexes of the inhabitants directly on the capsules. If after emergence the females are not used within a day or so, they will either chew out of the capsule or attempt to squeeze between the 2 halves. Maintaining the capsules in a foam- or cotton-stoppered vial will prevent wasps from gaining their freedom.

4. Constructing Observation Chambers

Suitable observation chambers can be easily and cheaply made by gluing a 5 mm tall ring of 12.5 mm diameter translucent tubing to a microscope slide. A glass or plastic coverslip provides a replaceable cover for confining the wasps during observation of courtship. Introduction of the wasps into the observation chamber is accomplished with a camel's hair brush, and is a skill that may require a bit of practice. Because males are unable to fly and generally move more slowly than females, add males to the observation chamber first.

5. Additional Hints

This exercise requires some skill and patience -- skill to handle the wasps gently and quickly, and patience to observe them until the courtship display begins. Because of their small size, their behavior must be viewed under a dissecting microscope. For optimal viewing, good lighting is important; placing an index card on the microscope stage provides a white background and enables the observer to gently move the observation chamber by smoothly sliding the card.

NOTE: If at all possible, microscope lamps should be provided with heat filters, as the wasps are very sensitive to heat and dessication. Alternatively, a container of water placed between the lamp and the microscope will serve as a crude but effective heat barrier.

The wasps are sometimes rather fickle in their behavior, and it is important to appreciate that individuals may vary greatly in their propensity to engage in courtship. For greatest success use virgin females about 2 days old. Male age is less critical. However, on occasion males have been observed to attack and eat 1 or 2 virgin females in succession prior to displaying courtship behavior. The reasons for this are not clear, but males that have been isolated for extended periods (greater than 2 days) seem especially prone to such cannibalistic acts.

Species level taxonomy of Melittobia is presently under study by E. C. Dahms of the Brisbane Museum, Queensland, Australia. No published keys for identifying the Melittobia species yet exist. However, it appears that at least 5 species commonly infest mud-dauber nests in North America. Exact species identification is not critical to the success of this exercise. However, because different species of Melittobia differ strikingly in the style and duration of courtship (see Evans and Matthews 1976, and van den Assem 1975), it is well to be alert to the possibility of having more than 1 species. (Different species may also differ in the ease with which they can be cultured on fly puparia.) Should behavioral or morphological differences indicate that you have more than 1 species in culture, the opportunity for undertaking comparative studies exists.

For background information on courtship in Melittobia, see Evans and Matthews (1976), van der Assem (1975), and van den Assem and Maeta (1978).

Additional References

Assem, J. van den. 1975. Temporal patterning of courtship behaviour in some parasitic Hymenoptera, with special reference to Melittobia acasta. J. Entomol. (A) 50: 137-146.

Assem, J. van den and Y. Maeta. 1978. Some observations on Melittobia species (Hymenoptera, Chalcidoidea, Eulophidae) collected in Japan. Kontyu 46: 264-272.

Cross, E. A., M. G. Stith and T. R. Bauman. 1975. Bionomics of the pipe-organ mud-dauber, Trypoxylon politum (Hymenoptera: Sphecoidea). Ann. Entomol. Soc. Am. 68: 901-916.

Evans, D. A. and R. W. Matthews. 1976. Comparative courtship behaviour in two species of the parasitic chalcid wasp Melittobia (Hymenoptera: Eulophidae). Anim. Behav. 24: 46-51.

25. FOOD-BASED TERRITORIALITY AND SEX DISCRIMINATION IN THE WATER
STRIDER, GERRIS. R. STIMSON WILCOX

1. Biology of G. remigis

In the northern United States and Canada, Gerris remigis overwinters as adults, which lie dormant amid debris along the water's
edge until early spring, when they become active on warmer days and
begin mating and egg laying. (For general ecology references on G.
remigis, see Brinkhurst, 1960; Calabrese, 1977; Galbraith and Fernando, 1977; Hungerford, 1919; Jamieson and Scudder, 1977; Matthey,
1975). Nymphs from these eggs reach adulthood by mid-June. Apparently depending on food levels, some male and female nymphs (from
third instar stage onward) may become territorial in some (but not
all) quieter pools and eddies, excluding other striders from areas of
around 20 to 40 cm in diameter. Territory boundaries are generally
defined by rocks, sticks, etc., along the edge, as well as by behavior itself. Adults (both sexes but mainly males) become territorial
within 1 to 2 weeks after the last molt if food levels in the stream
are low. Territoriality has not yet been observed during the spring-
time mating season, though it may sometimes occur then also (Wilcox,
in prep.; Wilcox and Ruckdeschel, in prep.). Some individuals of
summer-generation Gerris remigis (at least in the Northeast) can be
induced to become territorial in the laboratory simply by lowering
their food level, if the striders are in the proper laboratory
setting.

2. Identifying G. remigis

If keying out water striders is necessary, the following keys by
geographical region will be helpful:

Brooks, A.R. and L. A. Kelton. 1967. Aquatic and semiaquatic Heter-
    optera of Alberta, Saskatchewan, and Manitoba (Hemiptera). Mem.
    Entomol. Soc. Canada. no. 51. Ottawa. Ent. Soc. Canada.
Calabrese, D.M. 1974. Keys to the adults and nymphs of the species
    of Gerris fabricius occurring in Connecticut. Mem. Conn. Entom-
    ol. Soc. 1974: 227-266.
Cheng, K. and C.H. Fernando. 1970. The water-striders of Ontario
    (Heteroptera: Gerridae). Life Sci. Misc. Publ., R. Ont. Mus.
    Toronto.
Ellis, L.L. 1952. The aquatic Hemiptera of southeastern Louisiana
    (exclusive of the Corixidae). Amer. Midl. Nat. 48: 302-329.
Froeschner, R.C. 1962. Contributions to a synopsis of the Hemiptera
    of Missouri, part V. Amer. Midl. Nat. 67: 208-240.
Herring, J.L. 1950. The aquatic and semiaquatic Hemiptera of north-
    ern Florida. Part 1: Gerridae. Fla. Entomol. 33: 23-32.

Scudder, G.G.E. 1971. The Gerridae (Hemiptera) of British Columbia. Entomol. Soc. Brit. Colum. 68: 3-9.

Scudder, G.G.E. and G.S. Jamieson. 1972. The immature stages of Gerris (Hemiptera) in British Columbia. J. Entomol. Soc. Brit. Columbia 69: 72-79.

3. Laboratory Set-Up

An effective laboratory set-up can be made by obtaining or constructing a laboratory tray of probably no smaller dimensions than about 70 x 120 x 12 cm (Fig. 25-3 in exercise). The simplest way to produce the necessary surface current is to blow air from a pipette as shown. Alternatively, a water pump or propeller system can be used. Float a few thin flat pieces of polystrene here and there in the tank for resting places and possible territorial boundaries. The water will move in a large circle if large stones or bricks are placed more or less along the center and in positions that, along with the flat polystyrene pieces, form physical boundaries for slowly-flowing open water areas of about 20 cm in diameter (Fig. 25-3). These areas should become territories.

4. Prelaboratory Preparations

To set the stage for territoriality, for the minimum tank size of about 70 x 120 x 12 cm, add to the tank 5 male and 5 female nymphs (third instar stage or higher) or adults (or mix the 2 if you wish). Feed them excessively for at least 2 days. Water striders will feed on practically any small, soft arthropod, such as house flies, leafhoppers, etc. A convenient long-term laboratory source is flightless Drosophila, obtainable from most biological supply houses along with an instant medium for culturing them. (Several flightless Drosophila cultures begun 3 weeks before the lab, with new cultures begun each following week, would do the job well.) After the 2 days, clean the surface of all available food. Be sure that no more food can get into the tray. Then, noting the time, begin to let the striders go hungry.

5. Sex Discrimination Experiments

During territorial encounters, male nymphs (instars 3-5) and male adults produce high frequency (HF) ripple signals, as described in Wilcox (1979) and shown in Fig. 25-4 in the exercise. Both sexes in all stages also produce low-frequency signals. Magnetic tape (Magnet Tape, Jobmaster Corporation, Randallstown, MD) may be obtained from most local hardware stores.

NOTE: It would be advisable to try this experiment out yourself well before the students do, owing to the relative subtlety of making and putting on the masks, putting on the magnets, and timing the playbacks with the proper intensity of signal when the male is at an appropriate orientation and distance.

26. FOUNDRESS INTERACTIONS IN PAPER WASPS (Polistes). R. W. MATTHEWS

1. Species to Use and Optimal Season for Study

In the north central and eastern U. S. the most commonly encountered species nesting on buildings is Polistes fuscatus. According to Eberhard (1969; see exercise for complete references), this species often has multi-foundress nests. Noonan (1979) and Klahn (1979) have studied foundress behavior. In the southeastern and southern states the commonest wasp nesting on buildings is P. annularis, the largest species of Polistes in the U.S. This species is also typically pleometrotic. Hermann and Dirks (1975) and Strassman (1980) have discussed its spring behavior. In the south and southwest, P. exclamans also nests commonly on buildings. Its nests are more commonly, but not always, haplometrotic. In the far west P. apachus is found on buildings.

This exercise should be done during the time period between first egg-laying and first worker emergence, which in Georgia falls roughly between mid-April and mid-May. In New York and Michigan, P. fuscatus nests are at an optimal stage for these observations from late May to mid-June. Because foundresses and workers are not dimorphic, one cannot assume upon finding a nest with a few adults on it that these individuals are all foundresses. Instead, check to be sure that there are no opened pupal cells, indicative of worker emergence.

Keys to species of Polistes are difficult and the genus exhibits considerable variation within and between species. The key by Krispyn and Hermann (The Social Wasps of Georgia: Hornets, Yellowjackets and Polistine Paper Wasps. Univ. Georgia Agric. Exp. Sta. Res. Bull. 207, 39 pp., 1978) will permit identification of most species occurring east of the 100th meridian.

2. Finding and Marking Wasps

Nest location requires checking in advance of the exercise. Productive sites are abandoned buildings, sheds, picnic shelters, barns, etc. A cluster of buildings, as is often found on a farm, is usually productive enough to provide plenty of nests for a class of 20. Many nests will prove inaccessible except by ladder, but with diligence it is usually possible to locate enough nests favorably situated for closeup viewing without endangering life and limb. Students should be able to observe from a distance of no more than 1 meter, if possible.

Prior to the day of class, wasps must be individually marked. This should be done after dark with the aid of a flashlight with a piece of red cellophane over the light. Marking may be done several days in advance, preferably on a cool evening when the wasps will be somewhat sluggish. Testors enamel, typewriter correction fluid (such as Liquid Paper), or fast-drying acrylic paints may be used, and applied with a fine brush, toothpick or similar object almost anywhere on the thorax or abdomen of the resting wasps without having to capture or anesthetize them.

3. Sting Precautions

Foundresses of P. annularis and P. fuscatus are generally relatively docile and not easily provoked to attack. Although I have not had anyone stung while doing this exercise, the possibility of a sting incident always exists. Therefore, inquire in advance whether anyone knows that they are allergic to venom and take appropriate precautions.

In order to obtain meaningful data it is very important that the wasps not be disturbed. Stress this to the class repeatedly. Disturbed wasps will simply leave the nest for extended periods.

4. Observation Conditions

Because this is a field-oriented exercise, it is essential that data collection be done on a sunny afternoon (temperature above 22°C). Otherwise, students will observe very little activity, as wasps simply sit about and act very sluggish in cool overcast weather. Strive for at least 1 and ideally 2 hours of uninterrupted observation in order to generate a reasonable data base for analysis, prior to nest collection.

5. Parasites

There is a possibility that some cells in various nests will be parasitized. The presence of potential parasites in the nest vicinity is also possible. The instructor should be alert to these possibilities. Nelson (1969) provides information about the kinds of parasites attacking Polistes.

---

27. SOUND PRODUCTION AND COMMUNICATION IN SUBSOCIAL BEETLES. W.H. GOTWALD, JR.

Popilius disjunctus has never been successfully, i.e. continuously, cultured in the laboratory (Harrel, 1967). However, adults can be maintained in the laboratory with little or no care for several weeks, if not longer, when provided with rotting wood in the correct state of decay. Adults prefer decaying oak to other hardwoods and avoid pine, except on rare occasions. The adults can be kept in a covered plastic container with an adequate supply of decaying wood and a moistened paper towel to maintain humidity.

Adults of P. disjunctus have been collected from rotting logs at all times of the year, even during January and February in regions with rigorous winters (Gray, 1946). Thus, a dependable source of laboratory specimens may be at hand. This is especially true in the southeastern United States. However, if the collection of adults is inconvenient or impossible, they can be purchased from Carolina Biological Supply Company, Burlington, NC 27215. This supplier lists the beetle under its synonymized name, Passalus cornutus.

---

28. COMPARATIVE ANATOMY OF THE INSECT NEUROENDOCRINE SYSTEM. J.G. STOFFOLANO, JR. AND C.-M. YIN

A. Prelaboratory Preparations

1. The instructor should become confident in the dissections and locations of the structures discussed. Consult Cornwell (1968) for assistance with the cockroach, and Dethier (1976) and Cantwell et al. (1976) for the fly.

2. A fresh preparation of the methylene blue solution is preferable. It will, however, keep for at least 1 month and does not have to be refrigerated; when it turns blue, it should be discarded.

3. Dental wax can be ordered from Polyscience Inc., Warrington, PA.

B. Preparation of Methylene Blue Staining Solution

1. Add 0.4 gm methylene blue (U.S.P.) and 0.8 gm Rongalite (sodium formaldehydesulfoxylate) to 20 ml distilled water. Mix well and add 10 drops of 10% H_2SO_4.

2. Heat gently in a small water bath until the solution becomes clear greenish-yellow.

3. Filter and keep closed in vial for at least 24 hours. The solution should become clear light yellow (leuco-base). The leuco-base enters the cell and there is oxidized, thus staining it.

4. Just before use, dilute the light yellow colored methylene blue with 0.75% NaCl at the ratio of 1 to 4.

NOTE: Prepared slides showing the chemosensilla on the proboscis of the house fly can be purchased from most biological supply companies; however they do not show the internal nerves going to the sensilla.

---

29. GROOMING IN THE FLY, PHORMIA. W.G. EBERHARD AND J.R. MATTHEWS

---

1. Obtaining Flies: 2 Basic Alternatives

A. Visit a sizeable accumulation of garbage (e.g. behind a supermarket) armed with an insect net and numerous jars and vials. Captured flies can then be placed in a refrigerator for storage for a day or so before the lab (but not for extended periods); they are best kept with pieces of damp sponge or leaves to provide a humid atmosphere without giving them a chance to drown themselves. The disadvantages of this technique (except for comparative purposes) are that one usually captures a number of species, and many specimens must be discarded. In addition, one must determine which species will perform best after warming up from being chilled.

B. Raise flies on standard media and/or acquire them from supply houses. The genera commonly available are Sarcophaga, Calliphora, Musca, and Phormia. This method is preferable in that the condition of the flies prior to the experiment can be more or less equalized. Instructions for their care come with the purchase of the flies. Order the flies 3 to 4 weeks before intended use. The house fly, Musca, performs less reliably and its smaller size makes amputating its tarsi more difficult. If this exercise is performed with species other than Phormia, preliminary observations must be made to determine how the flies will perform after warming up from being chilled.

2. Prelaboratory Preparations

Have the flies ready and waiting in appropriate condition when the laboratory period starts, as the students will usually need most of the proposed 3 to 4 hours to complete their observations. Each student will need 2 warmed flies in clean vials and 1 chilled fly in a clean vial in an ice bath, plus 1 extra empty vial for transfers.

3. Notes on Laboratory Procedure

A relatively small amount of cornstarch in each vial is enough to ensure that the fly gets dusty. Normal movement within the vial is usually sufficient to dust the fly. However, you may recommend a gentle, quick shake of the fly within the vial if you prefer. This will help make the student aware of the imprecision in the application of cornstarch.

As vials, we recommend either plastic Drosophila culture vials with foam tops (available from major biological supply companies) or glass shell vials approximately 1 cm in diameter and 6 cm deep. Avoid vials with deep screw lids, as flies tend to move up inside the lid and become hidden from view.

---

30. LATERALITY IN DROSOPHILA. L. EHRMAN AND I.B. PERELLE

Flies used in this exercise should be taken from their respective stock bottles under light anesthesia shortly following eclosion -- no more than 4 hours after eclosing for D. willistoni, 8 hours for D. melanogaster and D. simulans, or 6 hours for D. persimilis and D. pseudoobscura. Males and females should be placed in separate vials, with moist food, to assure the virginity of the females and lack of sexual experience in males. At least 15 flies of each sex should be collected so that at least 12 male flies can be scored with 3 flies in reserve in the event some subjects are lost prior to scoring. Subjects should be aged 3 to 4 days (males preferably older).

---

31. CANDLES, MOTHS, AND ANTS: THE DIELECTRIC WAVEGUIDE THEORY OF INSECT OLFACTION. P.S. CALLAHAN

It is much easier to utilize wingless insects in this experiment than flying ones. There is some evidence from experiments run by a high school science fair researcher that cockroaches are also attracted to candles. If moths are used in this study, be sure to clip the wings about 0.5 mm back from the thorax.

Plastic, especially types of polyethylene, are extremely variable in their ability to pass far IR. Some thick black types block over 50% of the far IR radiation. For that reason it is best to try as many different types of plastic as possible. Comparisons of different plastics should prove interesting.

Caution students to treat the lighted candles with respect! Remember to put a piece of aluminum foil under the candle to prevent setting fire to the paper and table. Molten wax is a fire hazard.

EDITORS' NOTE: For additional background on rationale and procedure for statistical analyses, students may be referred to Exercise 2 (part C, descriptive statistics) and Exercise 29 (part F, Wilcoxon Test).

---

32. MODELING OF RANDOM SEARCH PATHS AND FORAGING PATHS. R. JANDER

---

1. The Structure of the Calculator Program

To enable the flexible and proper use of the calculator program (Fig. 32-1 in the exercise), its structure and the basic rationale behind it must be explained. The numbers in parentheses throughout this explanation refer to the listed program steps at which the respective calculations are carried out.

The central calculation cycle starts with the label A and the command to generate a random number that comes out of a range from zero to 1 (000-008). All the random numbers so produced from cycle to cycle are partitioned into 4 classes according to probabilities P_o, P_1, and P_2 which had been pre-entered into the registers 10, 11, and 12, respectively. (Note, 3 cuts make 4 pieces of cake.) Membership in these 4 classes is then translated into the numbers 0, 1, 2, and 3, which signify the turn angles (straight, left, back, right) of the simulation model and which are added to the absolute directions of the respective preceding steps stored in register 03 (036-044).

The result of this addition is the absolute direction of the last step expressed by the numbers 0-6. Since on the circle the directions 4, 5, and 6 are identical to the directions 0, 1, and 2 (see Fig. 32-2 in the exercise), 4 units are subtracted if the absolute angle is larger than 3 (045-061). The absolute direction of the last step so defined by 0, 1, 2, or 3 flashes briefly (062).

Next in the program, all steps in the 4 absolute directions (0, 1, 2, and 3) are separately counted and the counts are stored, respectively, in the registers 20, 21, 22, and 23 (063-069). At this point the central calculation cycle starts again with the generation of the next random number.

A certain number of cycles are combined into a run. The number of cycles per run is specified by the entry into the register 00 (070-073). During standard operation the number of cycles per run starts out with 2 and then increases geometrically from run to run: 2, 4, 8, 16, etc. (139-143, 077-083). The purpose of such geometrical increases is to have equal spacing of the sample points after their logarithmic transformation. At the end of each run the calculator is programmed to determine the endpoint of the whole path in terms of its rectangular coordinate values which are stored in the registers 20 (x-value) and 21 (y-value) (085-095). Rectangular-to-polar coordinate conversion is then employed to determine the linear distance of the endpoint of the path from the path origin (096-104).

2. Instructions to Program and Use the Calculator

The program (Fig. 32-1) has been written to be entered into either of the Texas Instruments calculators TI 58 or TI 59, containing in both cases the standard "master library". Familiarity with the basic use of these calculators, and the notations for their program listing, is presupposed.

After the program has been entered into the program memory, the following steps are necessary to carry out a random walk simulation:

1) Initialize the calculator by pressing the key E. This clears all the registers and prepares the cycle counter.

2) Enter an arbitrary seed number for the generation of the random numbers and press key E' (2nd E). The seed number is placed into register 09.

3) Enter the desired probabilities for the 0-, 1-, and 2-turn (straight, left, back) with keys A', B', and C', respectively. For the isotropic random walk, specifically, the sequence is: enter 0.25, press A'; enter 0.25, press B'; enter 0.25, press C'. For the random walk with persistence as discussed in the second activity, the sequence is: enter 0.9, press A'; enter 0.04, press B'; enter 0.02, press C'. The probability P_n for the right hand turn need not be entered. It is presumed by the program to be equal to the difference between 1 and the sum of the 3 entered probabilities P_0, P_1, and P_3.

4) Start the program execution by pressing the key A. Absolute directions flash, and the calculation stops at the end of a run, displaying the total number of steps (S) accumulated, that is, steps between origin and endpoint. Press the key R/S; the distance (D) between origin and endpoint is displayed for the previously displayed number of steps. Press R/S again, and then, after having executed the next run, again the accumulated number of steps (S) between origin and endpoint is displayed. Pressing R/S again produces the associated distance (D), and so on. The S and D values are all to be recorded and then used as discussed in the exercises.

As the runs become longer and longer during the assemblage of 1 random walk, more time is required for each subsequent run. As it takes about 4 seconds to execute the calculations necessary for 1 step, it is easy to estimate the overall calculation time for a longer random walk. Input and output numbers during program execution can be recorded automatically if the calculator is attached to a printer. In this case, print commands (prt) have to be placed at the program locations 076, 105, 114, 119, 127, and 136.

If you desire to have the absolute direction for every consecutive step displayed or printed (e.g. for graphing the path), then a R/S command, or a prt command, is to be placed in location 062.

3. Discussion of Square Root and Power Laws

The exponent p in the isotropic random flight simulation

probably turned out to be not much different from 0.5, which is the expected valueof it when a large number of repetitions are averaged. This then means that the function in question is a square root function:

$$D = kS^{.5} = k\sqrt{S}$$

(Remember, raising to the power of 0.5 is mathematically the same as taking the square root.)

Since this relationship holds for practically all isotropic random walks -- even those with randomly varying step lengths and those with turns varying continuously over 360 degrees -- one can speak of the square root law of the isotropic random walk or random flight.

After using the calculator-generated data points to determine the size of the exponent p for random flight with persistence, you will probably be surprised to discover that it also is not much larger than 0.5. In our particular simulation with strong persistence as well defined above, we can expect a p value of 0.56 with some random fluctuation up or down between repetitions of this exercise. Thus with moderate persistence we can still roughly apply the square root law. If we want a better approximation, we can simulate a random walk with a particular persistence in order to find the exponent for the above power function.

In conclusion, these exercises have familiarized students with important and unexpected properties of random walks, and have allowed them to "discover" the square root law and its generalization, the power law. This knowledge is not only important for the direct understanding of random foraging and search paths but has wider applications, too. For instance, one can use this knowledge to estimate flow of pollen carried by bees, or to find out whether released and recaptured marked Drosophila have dispersed by random flight.

---

33. SOUND COMMUNICATION IN AEDES MOSQUITOES: A "DRY LAB" EXERCISE
     IN DATA INTERPRETATION. J. R. MATTHEWS

This activity is intended as an introduction to data interpretation, particularly the aspects of selecting data pertinent to a problem and in selecting hypotheses which most adequately explain given data. The students are provided with only a very brief discussion of related aspects of the general biology of the mosquito, for the subject is covered in most introductory biology texts. If you wish to supplement this exercise with outside reading assignment, you might begin with Scientific American offprint no. 1106 (Jones, 1968). Roth's (1948) original study, while lengthy to assign as .

student reading, is well worth your own perusal as background for this exercise.

"ANSWERS" to Part B, Data Interpretation:

1. 1,4.
2. 5.
3. Points 5, 6, strengthened by morphological studies which actually have discovered a stridulating organ at the base of the female's wings.
4. There appears to be no specific sound produced by the female which can be interpreted as a "mating call", unless it be any sound she happens to make in flight, beginning several hours after emergence. The evidence that females continue to attract males by their buzzing even while they are actively attempting to repulse them argues against designating the sound a "mating call".
5. One advantage is that this behavior ensures the "attractiveness" of females which may produce sounds of slightly different frequencies because of certain uncontrollable factors such as size, wing damage, extent of distension abdomen with food, etc. A possible disadvantage might be that males may be attracted to females of other species and attempt to mate with them. However, since males mate repeatedly throughout their lifetime, this is less crucial than with some other insects.
6. In general, female mosquitoes must feed on blood to develop eggs. Pursuing only mature females increases the likelihood that the female has ingested a blood meal, and that mating will produce viable progeny. Newly emerged males generally remain quiescent until they complete their sexual maturation, thus behaviorally diminishing the frequency of their pursuit by mature males. By the time a male begins to fly voluntarily, the sound he makes in flight is sufficiently high in pitch to be beyond the range which will stimulate older males, and therefore he goes unmolested.
7. The observation (point 3 in the exercise) that mosquitoes will respond identically to a recording, a tuning fork, or an audio oscillator makes it unlikely.
8. A sure way to eradicate a species is to destroy its ability to reproduce -- thus, among the new methods of control or elimination of insect pests, such biological methods are particularly attractive. However, these require an intimate knowledge of the target species' way of life. With mosquitoes, what relatively little information is available for species other than Aedes aegypti indicates mainly that the sexual behavior varies remarkable from one species to another. Even in those instances where the sterile male technique has proved effective for mosquitoes in the laboratory, it has not met with much success under natural conditions. Unquestionably, male A. aegypti are attracted by certain sound frequencies and will continue to be attracted throughout their lifetime, but no experimental evidence has been found to indicate that females are attracted to males or to sound. Therefore, if sound is used as a means of attracting mosquitoes, all indications are that males would be the sex destroyed. It should be remembered that the bite, and the mosquito-borne diseases, are transmitted by the females. It is theoretically

possible that destroying a large percentage of males before mating could decrease the number of impregnated females to a point where the population became reduced to a medically unimportant level. However, those males surviving would still be in a constant mating state throughout their adult lives, capable of copulating repeatedly with the unreduced population of females.

---

34. DEVELOPING OBSERVATIONAL ABILITY AND GENERATING HYPOTHESES: NON-TRADITIONAL TRAINING EXERCISES FOR BEHAVIORAL SCIENTISTS. R.S. WILCOX

A. Alien Animals

This exercise rapidly and repeatedly tests the students' observational and hypothesis-generating abilities. Just as important, it makes the point -- as do all these rather unusual exercises -- that you can train yourself in ability as a behavioral scientist by practically any experience, whether everyday or unusual. This is also another way of noting that a genuinely inter-disciplinary mind perceives the interrelationships among all experiences, and integrates them into a characteristically flexible approach. Many recent authors have noted that highly creative, innovative minds are characterized by the interdisciplinary synthesis of what are normally regarded as unrelated ideas and information.

You might wish to discuss with students what factors seem involved in determining the difficulty level of a classification such as the Alien Animals. (Generally, it appears to depend upon the number of defining attributes, the amount of irrelevant information or noise, and on the subtlety of the attribute.) What analogies can they draw to problems in behavioral study?

For the Alien Animals pictured in this exercise, the principal identifying attributes are:

Blips: linear legs
Shibbles: curved body shape, 2 coiled "feelers"
Sharps: curved body shape, shading, simple facial features
Killiwonks: 1 pair of wings, knobbed antennae, 6 legs
Snappyloos: bilaterally symmetrical, same height as first row
Nippybugs: teeth showing, shaded circles or unshaded trapezoids

B. Just Looking

You may wish to share the classic situation where city slicker visits country bumpkin and finds, to his surprise, that his friend is less bumpkin than countrified and keeps pointing out deer and other animals that are obvious to country-bred eyes but well nigh invisible to city-bred eyes. However, after living on the farm for a while, the urbanite finds that without particularly trying he gets better and better at picking out camouflaged animals he would have missed entirely at first. This activity is an intensified experience in this type of visual pattern analysis.

C. Stereoscopic Fusion

If the copies of the stereoscopic fusion article avaialable to you do not include the paper glasses necessary for viewing the figures, they can be easily constructed at little cost. Buy a roll each of transparent green and transparent red cellophane at an office supply store or discount house. In 3 x 5 cards, cut holes 2.5-3.8 mm in diameter for eyes. Glue the cellophane in place, green on the left, red on the right.

D. Petals Around a Rose

With 5 to 10 students gathered around a table top, (more students than this is unwieldly for discussion immediately following the game) toss 5 dice down all together so they scatter randomly. Tell the students they have to figure out the rule or rules to tell the number of petals around the rose. Caution them that if anyone figures it out, he or she should keep quiet so the others have the opportunity to continue working on the problem.

The rule is simple: a dot in the center is a "rose", and any dots around it are "petals". You simply count the total number of petals, then after a short mysterious pause (as if you perhaps making a calculation in your head), you say "There are ___ petals around the rose." Wait for a moment to let the shock of your announcement sink in, then gather up the dice and repeat.

After several throws, most of the students will begin to get frustrated as pet theories are repeatedly foiled. Gauge their frustration carefully (some frustration being healthy and effective for training people). When the time is right, throw out 4, not 5 dice; then 3, then 2, then 1, counting out the petals as you go. By the time you reach 3 or 2, most will have discovered the rule.

Perhaps the most crucial point is now at hand -- to get the students to share the ideas/hypotheses they had while trying to figure out the rule. Occasionally someone will get the rule very soon, usually making everyone else feel inadequate, especially those who don't grasp it until near the end, or even must be told the rule. You should hasten to reassure everyone that not getting the rule soon simply means their search imagery (usually quantitative or mathematical) was inappropriate, not that they are stupid. Reemphasize that quantitatively-minded scientists usually have a harder time with this game than humanists, and especially emphasize that the game is meant to point out that non-quantitative patterns are normally very important in behavioral research as well as quantitative patterns, and that they should try to increase their ability in both approaches.

Incidentally, there is a mathematical solution to this game, and it is not uncommon for people to figure it out during the game.

Appendix 1
Comparing Behavioral Samples with Common Nonparametric Statistical Tests

The purpose of behavioral research is to determine the acceptability of hypotheses derived from theory. This is accomplished by collecting empirical data and deciding whether a particular hypothesis is confirmed by those data. This decision is always based on the same objective procedure, which follows these steps:

1. <u>State the Null Hypothesis (H_o)</u>.

This is the hypothesis of no difference. The alternative to it (H_1 or H_A) is an operational statement of the experimenter's research hypothesis. One hopes to reject H_o so that H_1 can be accepted, for this supports the research hypothesis and its underlying theory.

How H_1 is stated depends on whether the theory predicts only that 2 groups will differ, or includes the direction of the expected difference. When the direction is specified, a "one-tailed test" is used, which is more powerful than a "two-tailed test" (see below). For example, one might compare feeding rates between male and female praying mantids (Exercise 2), using the hypotheses:

H_o: Male and female feeding rates are the same.
H_1: Male and female feeding rates are not the same.

This is a "two-tailed" hypothesis because "not the same" could be either "greater than" or "less than." Or one could use the hypotheses:

H_o: Male feeding rates are less than or equal to female feeding rates.
H_1: Male feeding rates are greater than female feeding rates.

Here the research hypothesis is "one-tailed" because it specifies only 1 alternative to H_o.

2. <u>Choose the Appropriate Statistical Test to Use</u>.

Statistical inference is an orderly means for drawing conclusions about a large number of events on the basis of observations collected on a sample of them. The choice of which statistical test to use is dictated by your experimental design, the nature of your data (including the manner in which your sample was drawn and the level of measurement involved), the assumptions you are able to make about the nature of the population from which the sample was drawn, and the type of comparison you are interested in. These points are covered in more detail later.

3. For the Test You Have Chosen, Find (or Assume) the Sampling Distribution Under H_o for a Chosen Significance Level and Sample Size.

The sampling distribution is the theoretical distribution of values which would occur if we randomly drew all possible samples of the same size from the same population. Because we want to know whether the data we collect are part of this population, we must know the probability of certain values of the statistic occurring under H_o if we are to make a statement about the probability of our observed values doing so. (Generally, this step involves consulting published tables for the statistical test you have chosen.)

The significance level (alpha, α) is the probability of mistakenly or falsely rejecting H_o. The larger alpha is, the more likely that H_o will be rejected when it is in fact true. The most commonly used values of alpha are .05 and .01.

Sample size (N) is important because increasing N generally increases the power of a test -- that is, it increases the probability of correctly rejecting H_o. (For more detail, see "type I and type II errors" in any statistics text.)

4. Use Your Sample Data to Compute an Observed Value for the Statistical Test.

Generally, each statistical test has its own mathematical formula based on certain assumptions and mathematical models. The difficulty of the computations involved differs widely, but nonparametric tests are usually easier to compute than are parametric tests (see below).

5. See Whether Your Observed Value is in the Region of Rejection.

The "region of rejection" is the set of values so extreme that the probability of rejecting a true H_o is equal to or less than alpha. With one-tailed tests, this region is entirely at 1 end (tail) of the sampling distribution, whereas with two-tailed tests it is split between both ends of the distribution.

When your observed value is equal to or less than alpha, H_o is rejected, allowing you to accept H_1. Such an observed value is called "significant" at the specified alpha level.

PARAMETRIC AND NONPARAMETRIC TESTS

Some statistical tests make many (sometimes stringent) assumptions about the nature of the larger population from which your sample was drawn. Since population values are called "parameters", these techniques are called parametric. As a minimum, parametric tests assume that a known distribution is a suitable model for the samples (e.g., normal distribution or Poisson series) and specify that parameters of this distribution (e.g., μ, σ^2, λ) are the same for the samples being tested. In addition, parametric tests may be applied only to data of interval or ratio measurement (when values on the scale of measurement bear an exact relationship to their distances along the scale). When the conditions for their proper use are met, parametric tests are very powerful. However, the calculations they involve are often laborious. Two commonly used parametric tests are the F max test and the Student's t test. For detailed information on their use, see a biostatistical text such as Sokal and Rohlf (1969) or Zar (1974).

Other statistical techniques do not require assumptions about the parent population, and thus result in conclusions which require fewer qualifying statements. These nonparametric tests may be used on data which consist only of frequency of membership in one of a number of subclasses (ordinal or nominal levels of measurement), a common occurrence in behavioral research. Most nonparametric tests are relatively quick and easy to compute. They may be used on small samples, and their power can be increased simply by increasing the size of N. For such reasons, nonparametric tests have found special popularity among behaviorists. Detailed information on non-parametric tests is available in Siegel (1956).

SELECTED REFERENCES

Siegel, S. 1956. Nonparametric Statistics for the Behavioral Sciences. McGraw-Hill Book Co., New York.
Sokal, R. R. and F. J. Rohlf. 1969. Biometry. W. H. Freeman and Co., San Francisco.
Zar, J. H. 1974. Biostatistical Analysis. Prentice-Hall, Inc. Englewood Cliffs, New Jersey.

A KEY TO CHOOSING A NONPARAMETRIC TEST FOR TWO SAMPLES*

A guide to some common nonparametric tests when you have 2 samples is given below. Refer to the section indicated for explanations how to set up and compute each test. Tables of critical values for most of these tests are found in Appendix 2. Additional help may be found in the biostatistical texts mentioned above.

To use this key, choose between the statements bearing the same letter. For whichever better describes your experimental situation, go on to the next lettered choice indicated.

---

*Multisample testing requires analysis of variance (Sokal and Rohlf, 1969: 175-216; Zar, 1974: 130-139) and multiple comparisons (Sokal and Rohlf, 1969; 235-246; Zar, 1974: 151-161).

A. Nominal Data: number of animals distributed in each of a number of mutually exclusive subclasses which cannot be ranked relative to one another (for example, no. of males vs. no. of females, or no. receiving treatments A vs. B). Often frequency data (counts). Observations also must be independent (generally means 1 observation per subject). . . . . B
A'. Ordinal Data: number of animals in each of a number of mutually exclusive subclasses which can be ranked (ordered) along a scale. Subclasses need not be directly comparable (for example, subclass 4 need not be twice as high, much, etc. as subclass 2), but one should be easily recognizable as larger, smaller, more disturbed, more responsive, etc. than another. Data often in form of ranks . . . . D

B. Sample size of 40 or more, and expected frequencies all larger than 5. C
B'. Sample size 20 to 40 and smallest expected frequency less than 5, or sample size less than 20. . . . 1.3. Fisher Exact Probability Test.

C. Data in form of frequencies with which 2 independent samples fall into 1 or the other of 2 (or more) possible categories. From the observed values, expected values are generated and tested to find whether they are significantly different from the situation if the 2 treatments in fact made no difference . . 1.1. Chi Square (2x2) Test.
C'. Data in form of frequencies with which individuals in 1 sample fall into 1 or the other of 2 (or more) possible categories. Expected values are supplied from a hypothetical distribution based on no correlation between the variables . . . .
. . . . 1.2. Chi Square One Sample Goodness of Fit Test.

D. Comparing 2 independent samples . . . . 1.4. Mann-Whitney U Test.
D'. Comparing 2 related samples . . . . E

E. Data in form of correlated samples (such as matched or paired subjects, or repeated measures on same subject); tests for significant differences in location of the sample data. . . . . . .
. . . . 1.5. Wilcoxon Matched-Pairs Signed-Ranks Test.
E'. Data in 2 rank orders (every subject can be assigned a rank on both X and Y); measures tendency of the 2 rank orders to be similar
. . . . 1.6. Kendall's Tau Test.

THE CHI SQUARE ONE SAMPLE GOODNESS OF FIT TEST

The goodness of fit test follows the same general rules as the square (2x2) test discussed in section 1.1; consult that section more detail. However, the expected values in the goodness of fit are supplied from a hypothetical distribution based on the null thesis, rather than being generated from the observed values. The ness of fit test is based on the observed frequencies in a single le, rather than 2 or more.

This test is used with independent data in the form of observed encies (number of occurrences) for a single sample. It compares number of subjects, objects, or responses observed in various ories with a hypothetical distribution based on the null esis (such as that in Appendix 2.1) and decides whether the 2 ncy distributions are significantly different. To use this your sample size should be relatively large (greater than 40), e expected frequencies should each be larger than 5.

ıre

the chi square (2x2) test, the basic steps are:

ermine the hypotheses of interest.

your data into table form. (Here your table will be a ple one:

	Categories		
	A	B	etc.
frequency(O)			
frequency(E)			

ted frequency for each cell is based on your H_o, usually treatment had no effect.

te the observed chi square value, using the same formula as

$$\chi^2 = \Sigma \frac{(O-E)^2}{E}$$

for each category of observation, find the difference he observed and expected frequencies, square this and divide by the expected frequency. Finally, add the each category together to get your calculated chi square

your calculated chi square value with the published chi for that hypothetical distribution mentioned above. To ropriate value, you must know the significance level and

1.1. THE CHI SQUARE (2x2) TEST

The chi square (2x2) test is probably the most popular and widely applied statistical test among behaviorists. It is best suited for sample sizes of 40 or more, and expected frequencies all larger than 5. For smaller samples, use the Fisher Exact Probability Test (section 1.3) instead.

The chi square (χ^2) distribution is the base for a whole family of statistical tests used when one's basic hypotheses specify the proportions or probabilities of a series of observations falling into several mutually exclusive groups. Two common chi square analyses are the chi square (2x2) test and the chi square goodness of fit (section 1.2). Chi square tests are used with data in the form of frequencies (number of occurrences), never percentages or proportions. Data also must be independent of each other, which generally precludes taking more than 1 measurement from a single individual.

The chi square (2x2) test gets its name from the 2x2 "contingency table" which is set up to classify individuals in 2 directions with 2 categories in each classification. A "contingency" is simply a possible event. The samples might be any 2 independent groups (treated and controls, males and females, etc.). The categories might be any 2 mutually exclusive classifications or attributes (successes and failures, yes and no, above and below median, locations A and B, etc.). The test determines whether the 2 sample groups differ in the proportion with which they fall into the 2 classifications.

To do this, a chi square statistic is computed from the contingency table, and compared with the hypothetical distribution of values (Appendix 2.1) expected if the treatments made no difference.

Procedure

A. Determine the hypotheses of interest.

B. Set up a contingency table.

Group	Categories: no. of successes	no. of failures	
I	a	b	a+b=
II	c	d	c+d=
	a+c=	c+d=	a+b+c+d=n=

C. Calculate the chi square statistic.

The basic formula for chi square for a 2x2 table is

$$\chi^2 = \Sigma \frac{((O-E) - 0.5)^2}{E}$$

where O is the observed value for a given cell, and E is the expected value. The 0.5 is a "Yate's correction factor" which improves the accuracy of the chi square for a 2x2 table; it is omitted when larger contingency tables are used (see note below).

Calculating the chi square statistic from this formula, you would first find the expected value for each of the 4 cells separately, using the formula

expected frequency (E) = $\frac{\text{row total x column total}}{\text{grand total}}$

Then you would find the difference between observed and expected values for each cell, and subtract 0.5 from each cell for the correction factor. Each of these values would be divided by the expected values for that cell, and all the numbers for all 4 cells would be summed. An advantage of this long method is that you are able to check whether your expected frequencies are indeed all larger than 5, as is required for the most appropriate use of the chi square method.

There is a faster shortcut method, however. The formula looks complex, but is easier to compute than it appears:

$$\chi^2 = \frac{n(|ad-bc| - n/2)^2}{(a+c)(b+d)(a+b)(c+d)}$$

(The vertical lines indicate absolute value.)

D. Compare your calculated chi square value with published theoretical values, and reach a conclusion.

Use a table that lists the theoretical distribution of chi square under the hypothesis of no differences, such as that in Appendix 2.1. Such tables show how often a particular chi square value would be expected to occur by chance alone. (To find the proper place in the table, you must know the "degrees of freedom", which for a 2x2 table always equal 1.)

As you can see, your calculated statistic becomes larger in value with greater observed differences. At some point, it becomes so large it would be expected to occur only 5% (.05), 1% (.01), or even 0.1% (.001) of the time by chance alone. When your value is larger than the chance value at these levels, by convention we state that there is "a statistically significant difference between the 2

treatments" at that level. That is, the results are to have occurred by sampling variation alone.

NOTE: Larger contingency tables can be analyzed in same way as 2x2 tables. For these larger ta correction factor is not necessary, so the formula

$$\chi^2 = \Sigma \frac{(O-E)^2}{E}$$

where O is the observed value for each cell an column total/ grand total, as before.

To find the degrees of freedom for a large use this formula:
d.f. = (no. of rows - 1) x (no. of col

Example of Chi Square (2x2) Test Use

A behaviorist took several bottom samples stream, and counted the dragonfly naiads in ea species. Her null hypothesis was that the species she found in the 2 sample locations a 2x2 contingency table, her data looked like

Location	Species A
lake	a=42
stream	c=12
	a+c=54

Using the shortcut formula, she calcu

$$\chi^2 = \frac{n(|ad-bc| - n/2)}{(a+b)(c+d)(a+c)(}$$

$$= \frac{254(|(42)(104)}{(138)(116)}$$

$$= \frac{2,423,647,934}{172,886,400}$$

Comparing this value with the theoret freedom, the behaviorist found that point (10.83), so she was able to st sampling location and species was hig p < 0.001) level.

degrees of freedom. The significance level typically used is 0.05. The degrees of freedom is the number of categories minus 1.

E. A calculated chi square value larger than the tabular chi square value indicates that there is a significant association (at the probability level listed) between the variables.

Example of Chi Square One Sample Goodness of Fit Test

A graduate student forced crickets to turn left in a maze, then gave them a choice of turn directions. He wished to see whether the crickets' next turn was likely to be preferentially to the left or right. If the forced turn treatment had no effect, he reasoned, their next turn would be equally likely to be in either direction.

Set into tabular form, his data looked like this:

Turn direction after
forced left turn:

	Right turn	Left turn
Observed frequency (0)	38	12
Expected frequency based on equal distribution (E)	25	25
$(0-E)^2$	$(13)^2=169$	$(-13)^2=169$

Calculating chi square, he obtained the value:

$$\chi^2 = \Sigma \frac{(0-E)^2}{E} = \frac{169}{25} + \frac{169}{25} = \frac{338}{25} = 13.5$$

From a table of critical values of chi square (Appendix 2.1), using 1 d.f., he found that his value exceeded the tabular values even at the 0.001 significance level. Therefore he concluded that crickets were not equally likely to turn left or right after a forced left turn.

1.3. THE FISHER EXACT PROBABILITY TEST

The Fisher Exact Probability Test is a particularly useful nonparametric test for small, independent samples measured at the nominal or ordinal level. In its use and rationale, it is much like a chi square test.

The Fisher Exact Probability Test is used when the scores from 2 independent random samples all fall into 1 or the other of 2 mutually exclusive classes. As with the chi square (2x2) test, a contingency table is set up. The test determines whether the frequencies of the classes for the 2 groups differ from the situation that would occur if the frequencies of observations in the rows and columns were independent of one another.

Procedure

A. Determine the hypotheses of interest.

B. Arrange the data in a contingency table.

Set up a table as in section 1.1, with 1 variable being the rows and the other the columns. Calculate the marginal totals as indicated.

Sample, sex, species, etc.	Class, attribute, etc.		
I	(block A)	(Block B)	A+B =
II	(block C)	(block D)	C+D =
	A+C =	B+D =	n=A+B+C+D =

C. Calculate or find appropriate probability values.

After marginal totals are calculated, the appropriate values may be put into a formula to calculate an exact probability value for your data, to compare with theoretical values. If you have a calculator that computes factorials (or consult Appendix 2.8), the mathematics are considerably simplified:

$$P = \frac{(A + B)!(C + D)!(A + C)!(B + D)!}{N!A!B!C!D!}$$

However, an easier alternative is to simply consult published tables and directly determine the significance of your observed values without any tedious computations. (Consult a text such as Siegel, 1956, for the tables, which are too lengthy to include here.)

After finding your place in the table based on your (A+B) and (C+D) values, you will find several possible values of B listed. Find the observed value of B among these possibilities. (If the observed value of B is not among them, use the observed value of A instead, and use C in place of D.) Compare your observed D value with the values given under the various levels of significance on the table. If your D is equal to or less than the tabular value, then the observed data are significant at that level.

Example of Use of Fisher Exact Probability Test

A researcher studying water bug egg cannibalism found that 2 species appeared to differ in their tendency to eat their eggs. But were these differences due to sampling error alone? Because her sample size was small, she chose to run a Fisher Exact Probability Test rather than a chi square test.

In a 2x2 contingency table, her data looked like this:

	Adults cannibalistic?		
	yes	no	
species 1	3(A)	7(B)	A+B=10
species 2	8(C)	2(D)	C+D=10
	A+C=11	B+D=9	n=20

He consulted a published table, finding his place based on the (A+B) and (C+D) values, and discovered that his D was greater than the tabular value at p = 0.05. Therefore, he concluded that there was a significant difference between egg cannibalism frequency in the two species at the 5% significance level.

1.4. THE MANN-WHITNEY U TEST

The Mann-Whitney U test, one of the most powerful nonparametric tests, is used to determine whether 2 independent groups have been drawn from the same population. It is the nonparametric counterpart of the Student's t test, the most frequently used parametric statistical test for comparing means.

The Mann-Whitney U test is used to compare 2 independent samples when the data are measured at (at least) the ordinal level. Often the data are in the form of ranks. The samples must be drawn at random from 2 populations or arise from assignments at random of 2 treatments to the members of the same sample. (Samples need not be the same size.)

Procedure

A. Determine the hypotheses of interest, and whether the test will be one- or two-tailed (see section 1.1).

B. Place your data in the measurements columns of a table like the one below; the labeling of the groups as I and II is purely arbitrary.

Measurements for Group I (n_1 cases)	Overall Rank	Measurements for Group II (n_2 cases)	Overall Rank
	R_1=sum of ranks for Group I		R_2=sum of ranks for Group II

C. Determine overall ranks.
Consider all the measurements in both groups as one single sample, and rank all the measurements together (either from the highest to lowest or from the lowest to highest). Enter these ranks in the overall rank columns. For tied ranks, assign the mean of the ranks they would otherwise receive (see section 1.5, part C.2).

D. Focus on one of the groups. (If samples are of different sizes, choosing the smaller one is most convenient.) The value of U, the test statistic, is given by the number of times a measurement in the other group precedes a measurement in the group you are focusing on.

This can be determined directly by counting, but especially when samples are large it is less tediously obtained by the formula:

$$U = n_1 n_2 + \frac{n_1(n_1 + 1)}{2} - R_1$$

where n_1 = number of observations in sample I
n_2 = number of observations in sample II
R_1 = sum of the ranks of the observations in sample I.

E. Compare your observed U value with published values (such as those in Appendix 2.4) based on H_0. Since, as we noted, the labeling of the samples as I and II is purely arbitrary, it is possible to obtain 2 different values for U. A given table may require using either the larger or the smaller of them, so we need to know them both. However, this is not difficult since

$$U' = n_1 n_2 - U$$

F. Having compared the appropriate observed U with the critical value, reach a decision on statistical significance. Take care to use the right case (1- or 2-tailed) for your experimental hypothesis.

Example of Mann-Whitney Test Use

A behaviorist studying two samples of crickets suspected that the populations differed in song duration. Because he felt the variance in the populations was probably quite different, he was hesitant to use the Student's t test, and chose instead to run a Mann-Whitney test.

After ranking both groups from low to high, his data looked like this:

Song duration (secs) in population I	Overall Rank	Song duration (secs) in population II	Overall Rank
4	3	8	7
5	4.5	11	10
3	2	6	6
5	4.5	9	8
$n_1=4$	$R_1=14$	$n_2=5$	$R_2=32$

He calculated an observed U and U' for his data:

$$U = (4 \times 5) + 4(4+1)/2 - 14 = 16$$
$$U' = (4 \times 5) - 16 = 4$$

Comparing his U and U' with tabular values, he found that he could not reject H_0. Thus he concluded that cricket song duration in his two populations did not differ significantly.

1.5.   THE WILCOXON MATCHED-PAIRS SIGNED-RANKS TEST

Choose the Wilcoxon test when your study employs 2 related samples and yields difference scores which may be ranked in order of absolute magnitude.

This nonparametric test is very useful for the behaviorist wishing to test for significant differences when the samples are correlated (matched or paired).   One must be able to make the judgment "greater than" between any pair's 2 performances, and also between any 2 difference scores arising from any 2 pairs.

This test looks at the differences between 2 sets of measures on an individual, and compares these differences between 1 individual and another.   It gives more weight to a pair that shows a large difference between the 2 conditions, and less weight to a pair that shows a small difference.   Pairs that show no difference at all are simply ignored (dropped from the analysis).   The rationale behind the statistic, T, on which the test is based, is that if the 2 treatments that give rise to the 2 sets of measures on an individual are equivalent, then one should expect some of the larger differences to favor 1 treatment and some to favor the other.   In effect, the differences would cancel one another out.

Procedure

A.   Determine the hypotheses of interest.

The usual H_0 is that treatments A and B are equivalent.   The H_1 determines whether the test is one- or two-tailed (see Sec. 1.1). For the former, the researcher generally must predict in advance the direction of the difference (i.e., the sign of the smaller sum of ranks).

B.   List your data in the first 3 columns of a table like that given on the next page, so that pairs of observations are matched.   (The assigning of positive and negative values to the treatments is arbitrary; which treatment column you place on the right or left does not matter for the test.)

C.   Follow the steps below to fill in columns (1), (2), and (3):

1.   For each pair of scores, indicate in column (1) the numerical difference and whether this difference is positive or negative.

2.   Ignoring the positive or negative signs temporarily, rank the difference scores (d's) by their absolute values and enter them in column 2.   The smallest difference gets rank no. 1.   If a

Observational Unit (such as individual)	Treatment A(+) (B-)	(1) Signed Differences	(2) Rank of Differences	(3) Rank with less frequent sign
a				
b				
c				
etc.				

particular pair of scores shows no difference between the 2 treatments, drop it from the analysis. To score ties: if 2 or more d's are the same size, assign the average of the ranks that each would have been assigned if they had differed slightly. For example, if 2 pairs had d's of 1, pretend that one ranked first and the other second, and give each a rank of 1.5. The next largest d would be ranked third.

    3. Examine the positive and negative signs in column (1) and decide which is less frequent. In column (3), enter only the ranks from column (2) that correspond to the differences bearing this sign.

    4. Sum up the ranks in column (3) to obtain T, the statistic upon which the Wilcoxon test is based. In other words,

        T = the smaller of the sums of the like-signed ranks.

D. Compare your observed T with the appropriate published theoretical value based on H_o (see Appendix 2.5). To determine N, and thus which line on the table to consult, count the total number of different scores after subtracting any pairs that showed no difference between treatments. If your calculated T is equal to or less than the tabular value for a particular significance level and a particular N, H_o may be rejected at that level, and you may say there is a significant difference between the 2 groups of scores. (Remember that the significance level is determined by whether your hypothesis is one- or two-tailed, which depends in turn upon your experimental hypothesis; see section A, above.)

The procedure given here assumes 25 pairs of scores or less; for a larger sample, a slightly different technique is needed (see Siegel, 1956, p. 82).

Example Using the Wilcoxon Matched-Pairs Signed-Ranks Test

    A bright undergraduate student wondered whether "experience" affected cockroach mating. He watched a number of individually marked roaches during their first mating, and found a great deal of variability in the receptivity ("willingness") of females. He decided to represent this receptivity on a numerical scale of 1 to 10 for each female cockroach. Then he put the same roaches together a second time, keeping track of each and scoring the females' behavior as before. The student was confident that a higher score meant a more "willing" cockroach. He also knew that the difference between a score of 3 and 4, for example, was less than the difference between a score of 4 and 7. However, he was reluctant to say that the scale was sufficiently exact so that a difference between 3 and 4 was, for example, exactly the same as a difference between 5 and 6. Therefore, he decided to use a statistical test that would treat the scores as ranks rather than as numerically exact measures.

Choosing the Wilcoxon test, the student set up his data to look like this:

Female roach	Receptivity Score		d (+ or -)	Rank of d ignoring sign	Rank with less frequent sign
	at first mating	at second mating			
a	8	10	-2	2.5*	2
b	7	4	+3	4	
c	7	7	0**	---	
d	4	3	+1	1	
e	7	3	+4	5	
f	8	2	+6	6	
g	6	8	+2	2.5*	
h	10	2	+8	7	

T=2

*tied d's, so average rank assigned.
**no difference, so dropped from analysis.
N= 7 - 1 = 6. Rejection region, determined by H_1, 2-tailed.

Comparing his T with the published sampling distribution for T under the null hypothesis, the student found that for an N of 6, his value of 2 allowed him to reject the null hypothesis at alpha = .05 for a two-tailed test. Therefore, he could state that female cockroach receptivity for the first and second mating differed significantly at the .05 probability level.

1.6. KENDALL'S TAU TEST

This test is used when one wants to consider the linear relationship between 2 variables, neither of which is assumed to be functionally dependent upon the other. It requires at least ordinal measurement of both variables, so that ranking each variable is possible. However, it does not require that both variables in the comparison must be normally distributed, as does its parametric counterpart, the Pearson product-moment correlation coefficient.

In most of the statistical tests discussed in this appendix, we have been concerned with 2 variables, 1 of which was dependent upon the other. Sometimes, however, we may be interested in 2 variables which appear to be correlated but are both independent (for example, intragroup similarities or trait clusters). If the 2 variables are unrelated in the population, what is the likelihood (probability) that we would obtain a sample correlation as large as the one we have?

Two nonparametric tests are often used somewhat interchangeably in this situation. One is the Spearman's rank correlation coefficient (rho or r_s). The other is Kendall's tau (τ). Kendall's tau has 2 advantages. First, it has a sampling distribution practically indistinguishable from a normal distribution for sample sizes as small as 9. Second, it is generalizable to a Partial Rank Correlation Coefficient which can be used to examine the influence of unknown factors on both variables (see Note, below).

Kendall's tau may be thought of as a measure of the disarray between 2 sets of ranks. Each subject's measurements are ranked separately for the 2 variables. Then after arbitrarily arranging 1 of these sets of ranks in the "correct" order, one asks how often the ranks in the other set appear in the "wrong" order. Kendall's tau is the ratio of this number of "wrong" ranks to the total number of pairs one would have if every possible pair of both variables had agreed in their rankings. A simplified way of calculating this is outlined below.

Procedure

A. Determine the hypotheses of interest. (H_o is usually that there is no correlation between the 2 variables.)

B. Rank your data separately for each variable. For tied scores, use the average of the tied ranks as before. (In the formula for tau, ties change the denominator slightly, but the effect is relatively small; see Siegel, pp. 217-219.)

Place your data into a table like that given below, then follow the outlined steps to complete the table. Remember that calling the variables X and Y does not indicate that one is independent, the other dependent.

Subject: (N = sum of these. Their identifying codes will appear in some scrambled order.)

Rank on variable X	1	2	3	4 etc.
Rank on variable Y, (These ranks will appear in some scrambled order unless there is perfect correlation)				
(1) Y's ranks which are larger than this one (+)				(Since the last rank has no ranks to its right, it has no number in this column.)
(2) Y's ranks which are smaller than this one (-)				
(3) differences between (+) and (-) for this subject				

(1). Temporarily ignore the ranks for variable X and look at the ranks for variable Y. Starting with the first number on the left, count the number of ranks to its right which are larger than this first number. Enter on line (1).

(2). For this same first rank number, count the number of ranks to its right which are smaller, and enter in line (2).

(3). Subtract the negative number from the positive number directly above it, and enter the difference in line (3).

(4). Repeat this procedure for each subject. Then sum all the differences in line (3) to obtain S, the numerator in the formula for calculating Kendall's tau.

(5). The denominator of the formula represents the maximum possible score, represented by $1/2N(N-1)$, where N = number of pairs of scores.

C. Calculate tau = $\dfrac{S}{1/2N(N-1)}$

This value represents the degree of relationship between the 2 sets of ranks. The size of your calculated tau is a measure of the strength of the correlation; a tau of 1.00 indicates a perfect correlation. Tau is positive when the variables increase together, and negative with an inverse relationship.

D. To test the significance of tau: since tau is a function of S, the sampling distributions of S and tau are identical and either might be tabled (Appendix 2.6A, B). Usually S is chosen because it is more convenient.
     Compare your calculated S with the tabular S. value (appendix 2.6B). Choose a significance level, such as α = .05. If the probability given in the table is equal to or less than this for your S and N values, you can conclude that your 2 sets of data are correlated.

NOTE: It is always possible that such a correlation is due to the association between each of the 2 variables with a third variable, rather than to any genuine relationship between the 2 variables themselves. A statistical method to approach this problem is to retest your data while eliminating the influence of the third variable. This may be done with the Kendall Partial Rank Correlation Coefficient; see Siegel, pp. 223-229.

Example Using Kendall's Tau

     An entomologist found a species of tropical wasp in which males had home ranges of different sizes. The wasps also varied in wing length, a convenient index of general body size. She obtained data on 4 wasps, which she assumed to constitute a random sample from the population. Was there any association between home range and wing length in that population? Her data looked like this:

wasp	A	B	C	D
home range	very large	large	medium	small
wing length	2.5mm	4mm	1.7mm	3mm

She had no reason to assume that the 2 variables were functionally dependent, nor that both variables were normally distributed. Although she was not able to measure home ranges exactly, she felt confident of her ability to rank them. Thus she chose to run the Kendall's tau test.

First she ranked her wasps separately for home range size, then for wing length. Then she rearranged her data into a table with home ranges ranked in natural order like this:

Subject:	wasp	A	B	C	D
Rank on Variable X	home range	1	2	3	4
Rank on Variable Y	wing length	3	1	2	4
(1) Y's ranks which are larger than this one (+)		+1	+2	+1	
(2) Y's ranks which are smaller than this one (-)		-2	0	0	
(3) differences between (+) and (-) for this subject		-1	+2	+1	

To the right of rank 3, rank 4 is larger and 1 and 2 are smaller, so (+1-2)=-1. To the right of rank 1, 2 and 4 are larger but 0 are smaller, so (+2-0)=+2. To the right of rank 2, rank 4 is larger and 0 is smaller, so (+1-0)=+1. (Nothing is to the right of rank 4, so no computation is done with this rank.) Summing these, $S = (-1) + (+2) + (+1) = +2$ which is the numerator for Kendall's tau.

The entomologist knew $N = 4$ wasps and $S = +2$, so she was able to compute tau:

$$tau = \frac{S}{1/2N(N-1)} = \frac{+2}{1/2(4)(4-1)} = +.33$$

Thus she found a degree of relationship of .33 between the two sets of ranks. It appeared that wing length and home range size were related. Because her tau value was positive, she knew that the relationship was a direct, not inverse, one. In her wasp population, greater wing lengths seemed related to greater home range size.

To test the significance of her results, the entomologist consulted a table of S values like that in Appendix 2.6B for $S = 2$ and $N = 4$. There was a .375 probability of getting a value of S as large as she did. Thus, she could not say that her results were significantly different from chance. In other words, her tau value for a sample of 4 gave no evidence that would lead her to conclude the value of tau was any different from zero in the whole population from which the sample came. This was not surprising, however, because her sample size was very small.

Appendix 2
Statistical Tables

APPENDICES 2.1, 2.2.   Abbreviated table of critical values of χ^2.

d.f.	$\alpha=0.9$	$\alpha=0.5$	$\alpha=0.2$	$\alpha=0.05$	$\alpha=0.01$	$\alpha=0.001$
1	.016	.455	1.642	3.841	6.635	10.827
2	.211	1.386	3.219	5.991	9.210	13.815
3	.584	2.366	4.642	7.815	11.345	16.268
4	1.064	3.367	5.989	9.488	13.277	18.465
5	1.610	4.351	7.289	11.070	15.086	20.517
6	2.204	5.348	8.558	12.592	16.812	22.457
7	2.833	6.346	9.803	14.067	18.475	24.322
8	3.490	7.344	11.303	15.507	20.090	26.125
9	4.168	8.343	12.242	16.919	21.666	24.877
10	4.865	9.342	13.442	18.307	23.209	29.588

d.f. = degrees of freedom = (no. of rows - 1) x (no. of columns - 1)

Values of χ^2 equal to or greater than those tabulated occur by chance less frequently than the indicated level of α (2-tailed).   More extensive tables are available in Siegel (1956) p. 249, or Zar (1974) pp. 408-411.

APPENDIX 2.3.   Critical values for Fisher Exact Probability Test are too lengthy to reproduce here;   see pp. 256 - 270 in Siegel (1956) or pp. 518 - 542 in Zar (1974) for these tables.

APPENDIX 2.4. Abbreviated table of critical values of the Mann-Whitney U test statistic.

$\alpha = 0.05$ (2-tailed), 0.025 (1-tailed)

n_1	n_2=2	3	4	5	6	7	8	9	10	11	12	13	14	15	16	17	18
2							16	18	20	22	23	25	27	29	31	32	34
3				15	17	20	22	25	27	30	32	35	37	40	42	45	47
4			16	19	22	25	28	32	35	38	41	44	47	50	53	57	60
5		15	19	23	27	30	34	38	42	46	49	53	57	61	65	68	72
6		17	22	27	31	36	40	44	49	53	58	62	67	71	75	80	84
7		20	25	30	36	41	46	51	56	61	66	71	76	81	86	91	96
8	16	22	28	34	40	46	51	57	63	69	74	80	86	91	97	102	108
9	18	25	32	38	44	51	57	64	70	76	82	89	95	101	107	114	120
10	20	27	35	42	49	56	63	70	77	84	91	97	104	111	118	125	132
11	22	30	38	46	53	61	69	76	84	91	99	106	114	121	129	136	143
12	23	32	41	49	58	66	74	82	91	99	107	115	123	131	139	147	155
13	25	35	44	53	62	71	80	89	97	106	115	124	132	141	149	158	167
14	27	37	47	57	67	76	86	95	104	114	123	132	141	151	160	169	178
15	29	40	50	61	71	81	91	101	111	121	131	141	151	161	170	180	190
16	31	42	53	65	75	86	97	107	118	129	139	149	169	179	181	191	202
17	32	45	57	68	80	91	102	114	125	136	147	158	169	180	191	202	213
18	34	47	60	72	84	96	108	120	132	143	155	167	178	190	202	213	225

$\alpha = 0.01$ (2-tailed), 0.005 (1-tailed)

n_1	n_2=2	3	4	5	6	7	8	9	10	11	12	13	14	15	16	17	18
2																	
3								27	30	33	35	38	41	43	46	49	52
4				24	28	31	35	38	42	45	49	52	55	59	62	66	
5			25	29	34	38	42	46	50	54	58	63	67	71	75	79	
6		24	29	34	39	44	49	54	59	63	68	73	78	83	87	92	
7		28	34	39	45	50	56	61	67	72	78	83	89	94	100	105	
8		31	38	44	50	57	63	69	75	81	87	94	100	106	112	118	
9	27	35	42	49	56	63	70	77	83	90	97	104	111	117	124	131	
10	30	38	46	54	61	69	77	84	92	99	106	114	121	129	136	143	
11	33	42	50	59	67	75	83	92	100	108	116	124	132	140	148	156	
12	35	45	54	63	72	81	90	99	108	117	125	134	143	151	160	169	
13	38	49	58	68	78	87	97	106	116	125	135	144	153	163	172	181	
14	41	52	63	73	83	94	104	114	124	134	144	154	164	174	184	194	
15	43	55	67	78	89	100	111	121	132	143	153	164	174	185	195	206	
16	46	59	71	83	94	106	117	129	140	151	163	174	185	196	207	218	
17	49	62	75	87	100	112	124	136	148	160	172	184	195	207	219	231	
18	52	66	79	92	105	118	131	143	156	169	181	194	206	218	231	243	

The values in the above table are derived, with permission of the publisher, from the extensive tables of Milton (1964, J. Amer. Statist. Assoc. 59: 925-934). More extensive tables are available in Zar, 1974, pp. 475-487.

APPENDIX 2.5. Critical values of T in the Wilcoxon matched-pairs signed-ranks test.

N	Level of significance for one-tailed test		
	.025	.01	.005
	Level of significance for two-tailed test		
	.05	.02	.01
6	0	---	---
7	2	0	---
8	4	2	0
9	6	3	2
10	8	5	3
11	11	7	5
12	14	10	7
13	17	13	10
14	21	16	13
15	25	20	16
16	30	24	20
17	35	28	23
18	40	33	28
19	46	38	32
20	52	43	38
21	59	49	43
22	66	56	49
23	73	62	55
24	81	69	61
25	89	77	68

Adapted from Table I of Wilcoxon, F. 1949. Some Rapid Approximate Statistical Procedures. New York: American Cyanamid Company, p.13, with permission of the publisher. More extensive tables can be found in Zar, 1974, pp. 488-489.

APPENDIX 2.6A. Significance of Kendall's tau.

N	tau	N	tau	N	tau	N	tau
5	.80	12	.43	19	.33	26	.27
6	.69	13	.41	20	.31	27	.26
7	.63	14	.39	21	.31	28	.25
8	.57	15	.37	22	.29	29	.25
9	.53	16	.35	23	.29	30	.25
10	.49	17	.35	24	.28		
11	.45	18	.33	25	.27		

N = number of pairs of scores. The tau values are the smallest values of tau significant at the 0.05 level for different values of N. If your calculated tau value is larger than the tabular value, you can conclude that there is a positive correlation between your 2 sets of data. Adapted from Anderson, B.F. 1966. The Psychology Experiment. Brooks-Cole, Belmont, Calif.

APPENDIX 2.6B. Probabilities associated with values as large as observed values of S in Kendall's tau when the null hypothesis of no correlation is true.

S	Values of N				S	Values of N		
	4	5	8	9		6	7	10
0	.625	.592	.548	.540	1	.500	.500	.500
2	.375	.408	.452	.460	3	.360	.386	.431
4	.167	.242	.360	.381	5	.235	.281	.364
6	.042	.117	.274	.306	7	.136	.191	.300
8		.042	.199	.238	9	.068	.119	.242
10		.0083	.138	.179	11	.028	.068	.190
12			.089	.130	13	.0083	.035	.146
14			.054	.090	15	.0014	.015	.108
16			.031	.060	17		.0054	.078
18			.016	.038	19		.0014	.054
20			.0071	.022	21		.00020	.036
22			.0028	.012	23			.023
24			.00087	.0063	25			.014
26			.00019	.0029	27			.0083
28			.000025	.0012	29			.0046
30				.00043	31			.0023
32				.00012	33			.0011
34				.000025	35			.00047

Adapted from M. G. Kendall, Rank Correlation Methods, 3rd ed.(London: Charles Griffin & Co., 1962), Appendix Table I, by permission of the publisher.

APPENDIX 2.7. Random numbers uniformly distributed from 0 to .999999.

0.678219	0.590112	0.114941	0.771640	0.701021
0.382187	0.284315	0.665374	0.949462	0.558331
0.419148	0.109296	0.579846	0.497813	0.556162
0.503857	0.917950	0.650629	0.661108	0.731896
0.743769	0.996022	0.179356	0.991764	0.524256
0.509625	0.654127	0.838787	0.073375	0.899666
0.594991	0.121196	0.419818	0.708515	0.619191
0.387502	0.925539	0.066727	0.860929	0.739535
0.889192	0.378049	0.117557	0.041108	0.949019
0.810791	0.939448	0.865679	0.572145	0.125017
0.875124	0.075581	0.036205	0.173571	0.139108
0.979078	0.563396	0.448814	0.973246	0.989959
0.404655	0.816069	0.452421	0.500293	0.555179
0.315147	0.314384	0.486214	0.746099	0.905360
0.038879	0.397276	0.905748	0.469127	0.407365
0.697283	0.109289	0.734168	0.535837	0.594221
0.414818	0.912290	0.435423	0.638842	0.739739
0.862058	0.235922	0.190583	0.266025	0.120894
0.990575	0.410501	0.280109	0.726748	0.671871
0.073463	0.231195	0.571626	0.194774	0.043815
0.782275	0.435309	0.737204	0.804720	0.848594
0.893222	0.095052	0.988629	0.769645	0.667600
0.200087	0.938355	0.437382	0.271590	0.451632
0.273227	0.156826	0.838415	0.407451	0.822143
0.783488	0.890285	0.107106	0.666266	0.421884
0.913296	0.729569	0.076343	0.330020	0.804024
0.081967	0.567722	0.204317	0.546759	0.183929
0.934238	0.453971	0.830251	0.824642	0.684594
0.694236	0.789597	0.241060	0.524070	0.743571
0.852472	0.50920	0.731418	0.742898	0.342813
0.929165	0.823678	0.122340	0.777310	0.305997
0.427745	0.713510	0.887637	0.717949	0.343783
0.733459	0.062046	0.502929	0.794868	0.121880
0.838695	0.180799	0.963620	0.484101	0.501255
0.342537	0.052203	0.404137	0.129186	0.555338
0.073518	0.903010	0.879469	0.645524	0.039150
0.828336	0.431606	0.633235	0.885858	0.925439
0.630508	0.980958	0.550675	0.006790	0.293966
0.256851	0.985473	0.118050	0.710547	0.225767
0.011403	0.782539	0.280347	0.570394	0.869673

APPENDIX 2.8.   Table of Factorials.

N	N!
0	1
1	1
2	2
3	6
4	24
5	120
6	720
7	5,040
8	40,320
9	362,880
10	3,628,800
11	39,916,800
12	479,001,600
13	6,227,020,800
14	87,178,291,200
15	1,307,674,368,000
16	20,922,789,888,000
17	355,687,428,096,000
18	6,402,373,705,728,000
19	121,645,100,408,832,000
20	2,432,902,008,176,640,000

Appendix 3
Supplemental Resources

Arthropod Species in Culture in the United States and Other Countries by W. A. Dickerson, et al., a joint effort of the Entomological Society of America, the International Organization for Biological Control and the U. S. Department of Agriculture, was published in 1979 by the Entomological Society of America, 4603 Calvert Rd., College Park, MD 20740. It is a comprehensive listing of approximately 1,000 culture colonies comprised of 480 species belonging to 109 families. These colonies are maintained at 206 facilities in the U.S. and in 17 other countries. The book is organized into 3 sections: 1. an alphabetical list of genera and species, arranged by taxonomic order and family; 2. colony information, including diets, publications, age of culture, etc.; and 3. a directory of contributors. Although compiled primarily to support and encourge the use of economically important species for basic research, it also serves as a source of information about materials, techniques and related aspects of rearing operations.

The Bulletin of the Entomological Society of Canada contains an annual compilation of that country's arthropod colonies from which cultures may be obtained. Consult these publications for details on how to obtain particular insect species which may not be readily available otherwise.

Older, but still useful sources of culturing information for various common insects are found in Needham, J. G., et al., Culture Methods for Invertebrate Animals, published originally in 1937 by Comstock Publishing Company, Ithaca, NY 14850 and reprinted by Dover Publications, Inc., New York, in 1959. The UFAW Handbook on the Care and Management of Laboratory Animals, 3rd. ed. (1967), published by E. & S. Livingstone Ltd., Edinburgh and London, is another source of information for some species.

Carolina Biological Supply Company, Burlington, NC 27215, has published a pamphlet "Living Arthropods in the Classroom" by R. L. Best, (Cat. No. 45-4400, $1.50), which contains detailed rearing information on species they offer for sale (about 21 insect species).

Frass, is a bi-annual newsletter published by the Insect Rearing Group, currently composed of over 500 scientists involved in insect rearing programs throughout the world. It includes a variety of useful information, from lists of recent rearing publications to diet ingredient sources and price lists. As of this writing it is available to those involved in this work. For information, write Dr. Ronald E. Wheeler, Chevron Chemical Company, 940 Hensley St., Richmond, CA 94804.

Two other books known to us to include a selection of insect behavior exercises are: Price and Stokes (eds.), (1975) Animal Behavior in Laboratory and Field, 2nd. ed. (W. H. Freeman and Co., San Francisco) and Hansell and Aitken (1977) Experimental Animal Behaviour (Blackie and Son, Ltd, London).

A catalog of films "PCR: Films and Video in the Behavioral Sciences", which includes many dealing with insect behavior is available from Audio Visual Services, Pennsylvania State University, Univ. Park, PA 16802. Films can be rented or purchased.

Index